MW00697445

RHS
Horti
Curious

Horti Curious: A Gardener's Miscellany of Fascinating Facts & Remarkable Plants
Author: Ann Treneman
First published in Great Britain in 2024
by Mitchell Beazley, an imprint of Octopus Publishing Group Ltd,
Carmelite House, 50 Victoria Embankment, London EC4Y 0DZ
www.octopusbooks.co.uk

An Hachette UK Company
www.hachette.co.uk

Published in association with the Royal Horticultural Society

ISBN: 978-1-7847-2963-9

A CIP record of this book is available from the British Library

Set in Rosha Keyline, Avenir LT Pro and Larken
Printed and bound in Bosnia and Herzegovina

Conceived, designed and produced by The Bright Press,
an imprint of the Quarto Group, 1 Triptych Place, London, SE1 9SH
www.quarto.com

Publisher: James Evans
Editorial Directors: Isheeta Mustafi, Anna Southgate
Managing Editor: Jacqui Sayers
Art Director: Emily Nazer
Senior Editor: Izzie Hewitt
Project Editor: Katie Crous
Cover and layout design and illustrations: Matt Windsor

Mitchell Beazley Publisher: Alison Starling
Mitchell Beazley Editorial Assistant: Ellen Sleath
RHS Publisher: Helen Griffin
RHS Head of Editorial: Tom Howard
RHS Books Editor: Simon Maughan

The publishers wish to thank the following experts for their reviews: Guy Barter, Jordan Bilsborrow, Helen Bostock, John David, Gemma Golding, Mark Gush, Alex Hankey, Hayley Jones, Marc Redmile Gordon.

The Royal Horticultural Society is the UK's leading gardening charity dedicated to advancing horticulture and promoting good gardening. Its charitable work includes providing expert advice and information in print, online and at its five major gardens and annual shows, training gardeners of every age, creating hands-on opportunities for children to grow plants and sharing research into plants, wildlife, wellbeing and environmental issues affecting gardeners.

For more information visit www.rhs.org.uk or call 020 3176 5800.

Horti Curious

A GARDENER'S MISCELLANY OF FASCINATING FACTS & REMARKABLE PLANTS

Ann Treneman

MITCHELL BEAZLEY

Lilium speciosum
'Imperiale'

Contents

Introduction

I am sure that all gardeners can remember the moment when plants became 'real' to them. For me, it was when I embarked on the project of creating the first garden that was mine alone. I say garden, but the space wasn't more than a metre long and half a metre wide, a rectangle of soil squished between the garage and the concrete path to the house. Small space, wide horizon: I wanted to grow herbs and, as I researched what to plant, a new world opened up to me of flavours and cures, both ancient and modern. Here, in my small rectangle, there was so much more to contemplate than just a few plants.

This is true about almost every aspect of horticulture: there is always much more than meets the eye. It has been such fun to be able to explore the wonderful kaleidoscope of facts, figures and foliage that make up this miscellany of what could be called 'plantology'. What a pleasure it has been to delve into the Horti Curious world in all its infinite variety and glory.

If plants could talk, they could tell us some amazing stories – of survival, derring-do, drama, beauty, friends and enemies. They are wild in many ways, and this book seeks to reach the parts that others have not. I think of it as a sort of cabinet of curiosities, or *wunderkammer*, which were all the rage in the Renaissance period. Except that instead of a wooden piece of furniture, the objects and stories are collected here in the pages of a book. The subjects are as small as an orchid seed and as large as Greenland, as old as Adam and Eve and as young as tomorrow. Together, they tell an endlessly surprising and intriguing story about a green and ever-changing world.

Ann Treneman

Rosa indica 'Fragrans'

Strelitzia reginae

Chapter 1

Beautiful Botany

This chapter is devoted to the wonderful world of botany in all its guises. It includes extraordinary plants, flowers that cure and kill, and the secret meaning of bouquets.

What is a flower?

Anatomy of a flower

We think of a flower as being a bloom, a thing of beauty, a blossom; but, by definition, a flower is where sexual reproduction takes place, the precursor to the seed or fruit. It's not all about sex, though, in that it also has organs that enclose and protect the emerging flower (sepals) and a ring of petals (corolla) to attract pollinators. The reproductive organs are stamens (male) and carpels (female). This is the basic anatomy, but within the realm of the flowering plants (Angiosperms), there is huge anatomical variation, such as unisex flowers, and those that are small and inconspicuous – grasses, for example, which are wind-pollinated and have no need to attract pollinators.

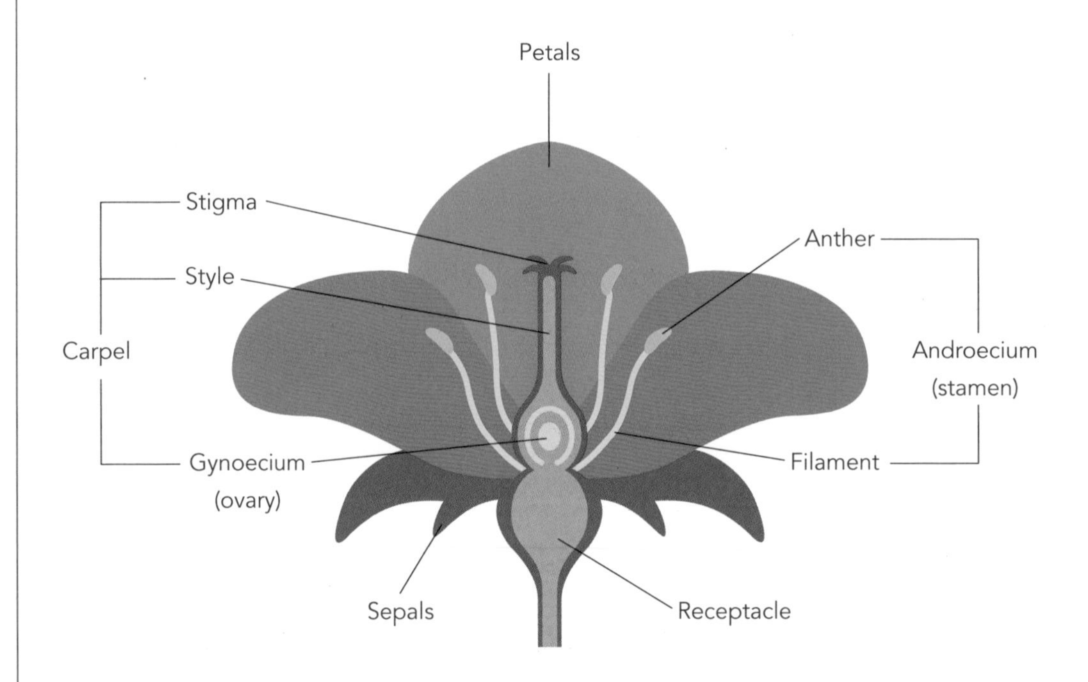

Rhododendron

Inflorescences

An inflorescence is how flowers are arranged on the floral axis of a plant. The aggregation or cluster of flowers is displayed in a specific pattern. There are two overall types: racemose and cymose. In the latter, the central flower opens first and the growth is continued by axillary buds. In the former, the axillary flowers form first.

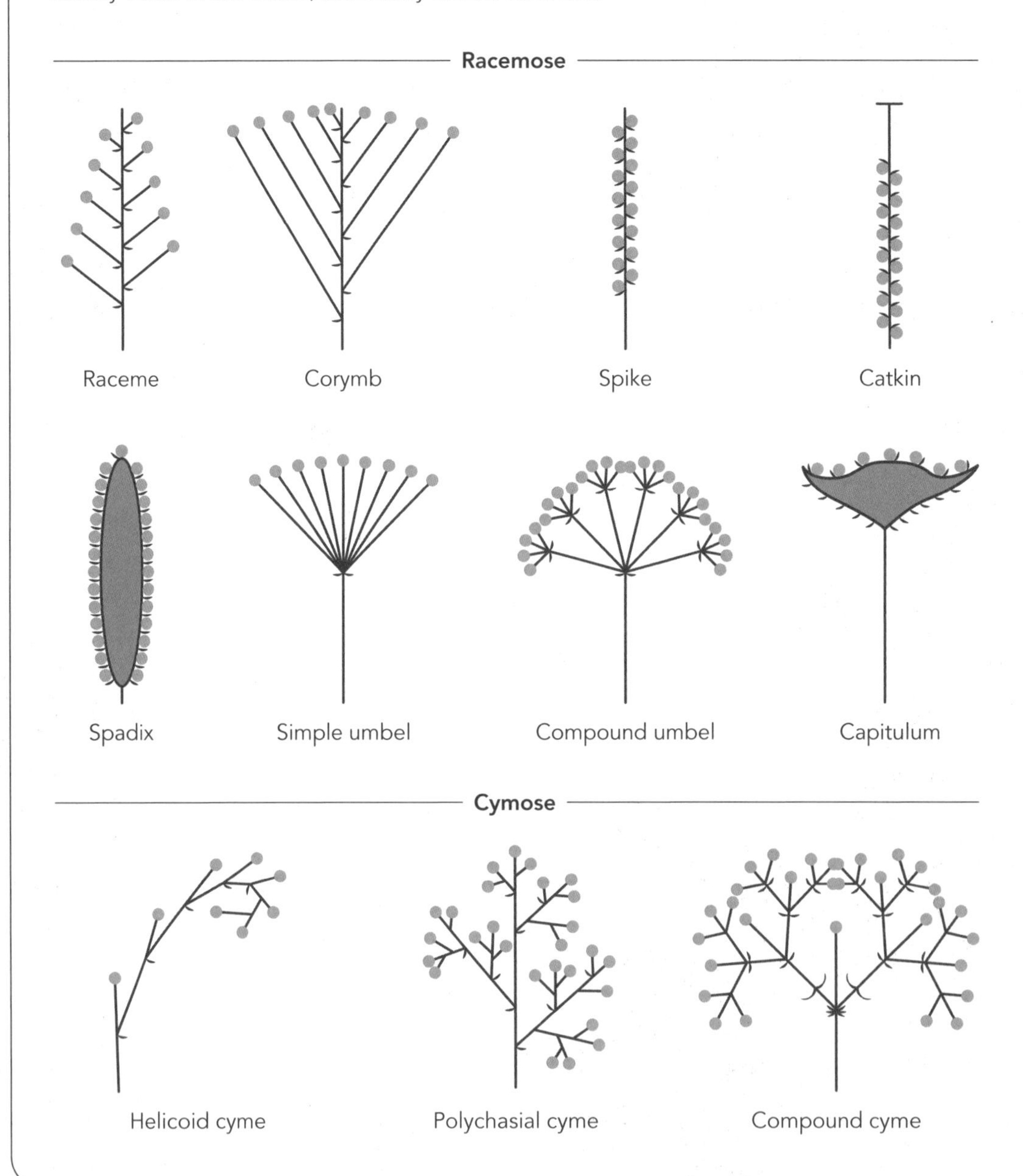

Blooming confusing

How many species of flower exist on Earth? When it comes to this question, you have to take into account some hefty 'known unknowns', not to mention the 'unknown unknowns'. The list of plants found and recorded is constantly changing, with the most recent Royal Botanic Gardens, Kew, State of the World's Plants and Fungi report in 2023 estimating that approximately 2,500 new species of all types of plants are identified yearly. It's not a 'known known' but it is generally agreed there are about 350,386 species of flower. It is also estimated that between 10 and 20 per cent of plants (though not all flowering) have yet to be found. The report estimates that 45 per cent of known flowering species are threatened with extinction. So, the best answer to the question of how many species of flower exist? It's somewhere north of 400,000. Probably.

The oldest flowers

The earliest flowering plants were not the showiest of blooms. It was 131 million years ago, during the early Cretaceous period: Earth was relatively warm, with forests that extended to the poles. The numerous inland seas were inhabited by marine reptiles while dinosaurs ruled on land. The first flowering plants were *Montsechia* (found in Spain) and *Archaefructus* (China). Neither had petals and both were aquatic. They looked something like hornwort (*Ceratophyllum demersum*), with squiggly stems and small leaves, but crucially they possessed reproductive organs. Their crown of 'oldest flower' may be under threat, however, as another plant fossil named *Nanjinganthus dendrostyla*, from China 50 million years earlier, is being studied as a possible winner.

Ceratophyllum demersum

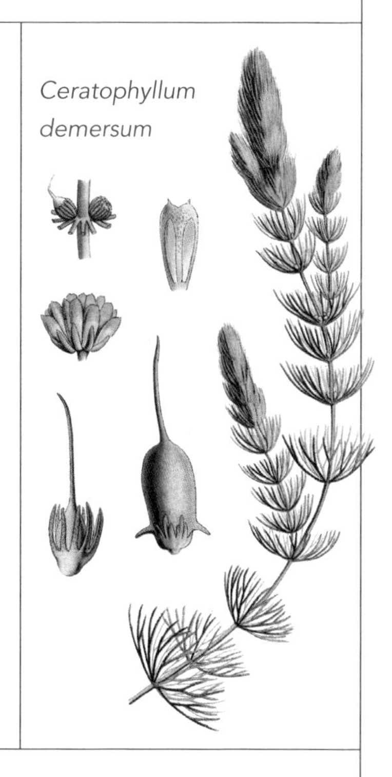

'By plucking her petals, you do not gather the beauty of the flower.'

Rabindranath Tagore, Bengali poet and polymath

The Doctrine of Signatures

Like cures like

The Doctrine of Signatures is based on the timeworn idea that plants can cure the parts of the body that they resemble. It's an ancient belief that has been found in traditional cultures around the world. The idea that 'like cures like' (*similia similibus curantur*) is a cornerstone of traditional Chinese medicine and Indigenous American herbalism and is also found in Asian, African, pre-Columbian and European cultures. The Greek physician and botanist Dioscorides explained in 65 CE how it works: 'The Herb Scorpius resembles the tail of the Scorpion, and is good against his biting.' It is not clear which herb he is referring to, though nettles are supposedly one of the Scorpion herbs.

Ginseng, the reluctant celebrity

The root of the shade-loving Asiatic ginseng (*Panax ginseng*) looks unimpressive in its naked state. It has a small but sturdy root 'body' with a tangle of smaller and more delicate roots attached to it. *Ginseng*, in Mandarin, translates as 'person root' or 'man essence', and *Panax* translates as 'cure all'. In traditional medicines, ginseng is valued as a tonic for the whole body: a general pick-me-up that has some added value as an aphrodisiac and elixir. The plant itself is a reluctant celebrity: it grows on the forest floor, is painfully slow to germinate (can take two years) and is also slow growing. It needs to be six years old to harvest. Asiatic ginseng is said to be almost extinct (and if found fetches huge prices) and has been replaced by 'wild simulated ginseng' or, the cheapest, a farmed variety. The American version, *Panax quinquefolius* (used by Indigenous Americans), is also over-harvested and is subject to some export controls.

Panax ginseng

Medicinal plants

Birthwort (*Aristolochia clematitis*): Twining plant whose flowers are thought to look like the womb and birth canal. Was used in Europe to treat women in childbirth; later found to be toxic to the kidneys (nephrotoxic).

Eyebright (*Euphrasia officinalis*): Small wildflower with white flowers that resemble the human eye. Was used to treat eye infections.

Heart's ease (*Viola tricolor*): The upper part of the flowers was thought to resemble the upper lobes of the heart. Was prescribed as a general heart tonic as well as for healing a broken heart.

Liverleaf (*Hepatica nobilis*): This small flowering plant has three-lobed leaves that someone with quite an imagination thought looked like the liver. Was used as a general liver tonic.

Delta maidenhair fern (*Adiantum raddianum*): A plant with a plethora of greenery that was considered (and still is by some) to be a cure for baldness and thinning hair, with extracts drunk or poured on the head.

Pomegranate (*Punica granatum*): The many-seeded fruit that is believed to resemble a human jaw and teeth. Was prescribed to cure toothache.

Viola tricolor

Punica granatum

Nomenclature: It's all in a name

The naming of names

There is a long and convoluted history of our attempts to find a scientific and orderly way of categorizing plants. For centuries the overall concept wasn't so much chaos as a hodgepodge of conflicting ideas. Should plants be organized by a single characteristic, such as the number of flower petals, as French botanist Joseph Pitton de Tournefort (1656–1708) thought? His English contemporary John Ray (1627–1705) advocated a more sophisticated method involving other attributes. A young Swede, Carl Linnaeus (1707–78), who had been obsessed with botany from the age of eight if not before, thought the best way was by sexual characteristics.

Linnaeus's sexual system, with or without bed curtains (see right), did not hold up over time (genetics won the day instead), but what has stayed, and has made him famous for all time, is the way he ordered the naming of plants. Gone were the long and wordy phrases, such as *Physalis annua ramosissima, ramis angulosis glabris, foliis dentato-serratis*, and in came the far snappier *Physalis angulata* (ground cherry).

This way of labelling plants, by genus first then species, via a system known as binomial nomenclature, was not new, but Linnaeus used it with a consistency that meant it caught on for good. Now everyone knew exactly which plant was being discussed. Linnaeus liked things 'neat', as he said. He was the son of a rector (and disappointed his father by not becoming one himself) and never lost his faith. One of his favourite sayings was: 'God created, Linnaeus organized'. This itself is, perhaps, the pithiest autobiography ever.

Sexual plants

Linnaeus first used sexual characteristics as a classification tool in his landmark book *Systema Naturae*, published in 12 volumes between 1735 and 1768, and which resembles an 18th-century version of an Excel spreadsheet. He was positively florid in his way of describing sex and plants: 'The actual petals of a flower contribute nothing to generation, serving only as the bridal bed which the great Creator has so gloriously prepared, adorned with such precious bed curtains, and perfumed with so many sweet scents, in order that the bridegroom and bride may therein celebrate their nuptials with the greater solemnity.'

The rules

Scientific names are based on an international code of nomenclature. The name always starts with the 'genus', which is the principal taxonomic category below 'family' and indicates a group that shares many structural similarities. 'Species', which is the next category, is even more specific about shared characteristics, and so on.

- Genus is written first and is capitalized.
- Species comes second.
- Both genus and species are italicized. So, the common wallflower illustrated right is *Erysimum cheiri*
- Subspecies are noted after this as: subsp. (or ssp.) *name*, as in *Erysimum candicum* subsp. *carpathum*
- Variety is noted as: var. *name*, as in *Erysimum capitatum* var. *purshii*
- Forma is noted after the species as: f. *name*, as in *Erysimum amurense* f. *flavum*
- Cultivar (from cultivated variety) is noted in single quotes, in roman, with capital first letters, as in *Erysimum* 'Bowles's Mauve'
- Hybrids are denoted by having an × (a multiplication sign, not the letter) which between the genus name and the species epithet, as in *Erysimum* × *marshallii*

Erysimum cheiri

A tree for a tree (and other plants)

Plant taxonomy is the science that deals with identifying, describing, classifying and naming plants. There are seven major categories of classification for plants, in descending order of shared characteristics. The groups can also be illustrated via plant family trees, or phylogenetic trees. These four examples show how members of the plant kingdom are categorized and the major ways they differ from one another.

Common pea: *Pisum sativum*
Kingdom: Plantae
Division: Embryophytes (land plants)
Class: Angiospermae (flowering plants)
Order: Fabales (legumes)
Family: Fabaceae (pea family)
Genus: *Pisum*
Species: *sativum*

Red bog moss: *Spaghnum capillifolium*
Kingdom: Plantae
Division: Bryophyta (moss and liverworts)
Class: Sphagnopsida (mosses)
Order: Sphagnales
Family: Sphagnaceae (*Spaghnum* mosses)
Genus: *Spaghnum*
Species: *capillifolium*

Soft tree fern: *Dicksonia antarctica*
Kingdom: Plantae
Division: Polypodiophyta (ferns)
Class: Polypodiopsida (vascular plants)
Order: Cyatheales (most tree ferns)
Family: Dicksoniaceae (tree ferns)
Genus: *Dicksonia*
Species: *antarctica*

Coastal redwood: *Sequoia sempervirens*
Kingdom: Plantae
Division: Pinophyta (conifers)
Class: Pinopsida
Order: Cupressales
Family: Cupressaceae (cypresses)
Genus: *Sequoia*
Species: *sempervirens*

Sequoiadendron giganteum

Out-of-this world plants

Begonia darthvaderiana: The very dark leaves of this tropical plant discovered in West Kalimantan in Indonesia in 2013 inspired C.W. Lin and C.I. Peng to go to the dark side for its name.

Mandevilla sherlockii: A new species of rock trumpet discovered in Mexico in 2017 and named after the great fictional detective Sherlock Holmes, because so much of taxonomy is detective work.

Stelis oscargrouchii: A small, delicate orchid found in Ecuador in 2015, which is a long way from Sesame Street, where a green puppet named Oscar the Grouch lives in a trash can and sings 'I Love Trash'.

Dracula smaug: An orchid found in Ecuador in 2015. The name 'Dracula' doesn't refer to anything to do with vampires but means 'little dragon'. 'Smaug' refers to the fictional dragon of Middle-Earth in *The Lord of the Rings* trilogy by J.R.R. Tolkien. So, this orchid is a double dragon.

The busy lives of leaves

Who'd be a leaf?

Leaves do not lead a relaxed existence. It's their job to bring home the bacon, or the plant equivalent, which means going about the business of photosynthesis, absorbing carbon dioxide and sunlight, combining this with water from the roots and turning it into glucose, and emitting oxygen as a waste product. It's an important balancing act that requires them to do this while being able to control heat intake and water loss, as well as minimizing the risk of freezing. Leaves are multitaskers (see Adaptations, opposite), vital to the health of the plant, whichever type they are or wherever they live.

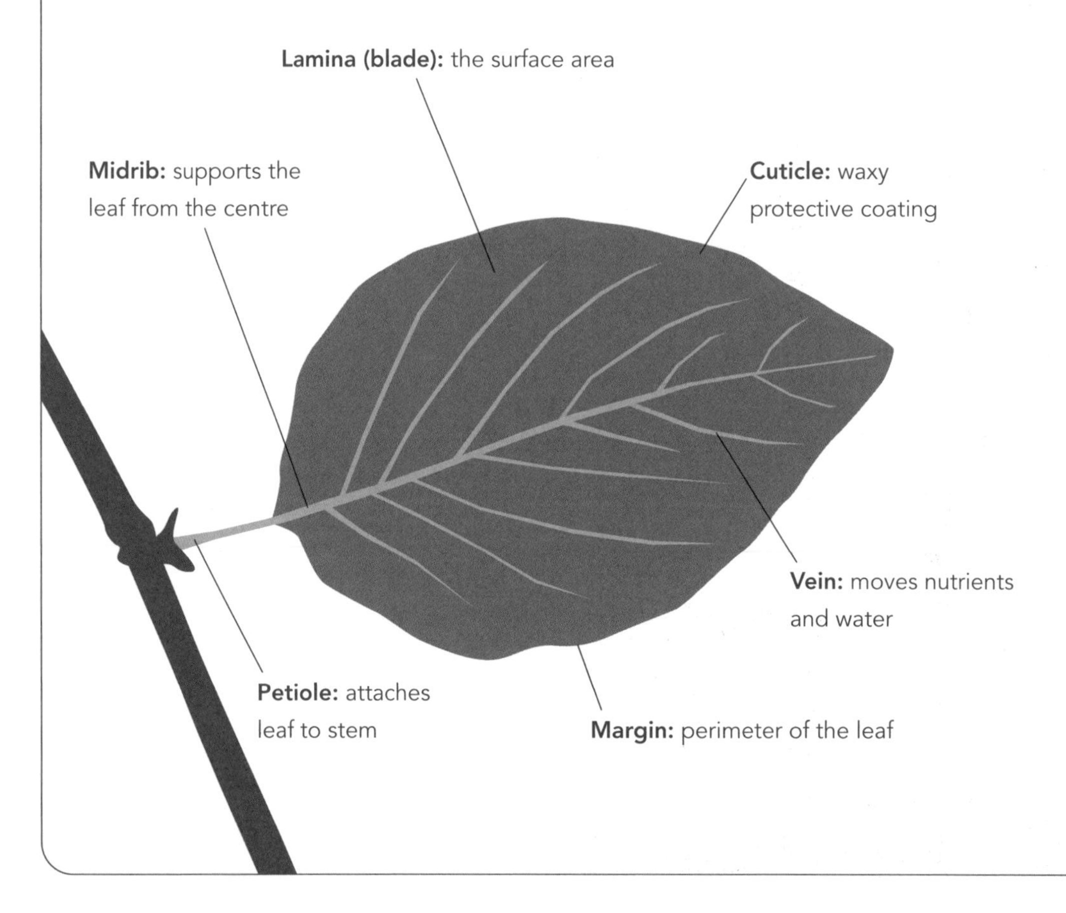

Adaptations for an easier life

- Plants such as cacti reduce water loss by keeping stomata (pores on the leaf surface) closed during the day. Pores open at night to allow gas exchange.
- Tropical forest leaves tend to be large and wide, as they are competing with so many other plants for sunlight and need plenty of surface area for transpiration (evaporation) to keep cooler.
- Leaf width is often narrower in drier climates to reduce surface area and minimize water loss.
- Alpines are compact with small leaves, which reduces the impact of frost damage.
- Needle-shaped leaves restrict water loss, allowing trees and plants to grow in dry conditions.
- Large, broad leaves are less common in coastal or exposed sites as they are prone to wind damage.
- Needle-shaped leaves have adapted to the cold by becoming thick, having a limited surface area and being coated in a waxy substance that traps moisture.

Shape-shifting for the future

The hop bush (*Dodonaea viscosa*) is an Australian native. It was traditionally used by Indigenous Australians to treat toothache, cuts and stingray stings, and by early European Australians to make beer. It has spread through much of the warmer parts of the world and, in Australia, it is being studied for its ability to change the shape of its leaves according to its environment. In adaptation to warmer, drier climates, its leaves become narrower. It's all about survival, and this ability to adapt will help plants respond to the changing climate.

Trithrinax brasiliensis, a tropical palm with large wide leaves

Leaf shapes and forms

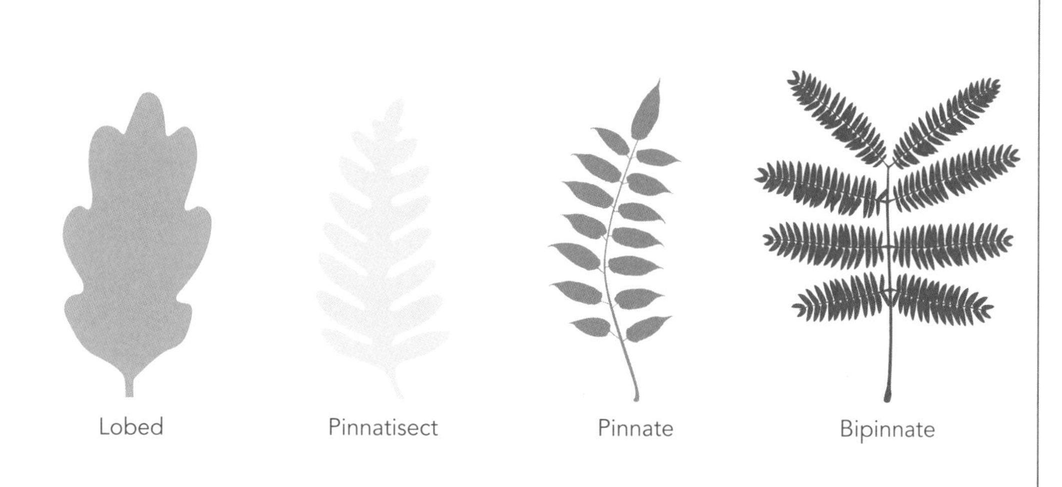

Record holders

The largest-ever leaf on record is a *Victoria boliviana* water lily pad that was measured at 3.2m (10ft 6in) in diameter. The smallest belong to the watermeal plants (*Wolffia*), with each leaf less than 1mm long.

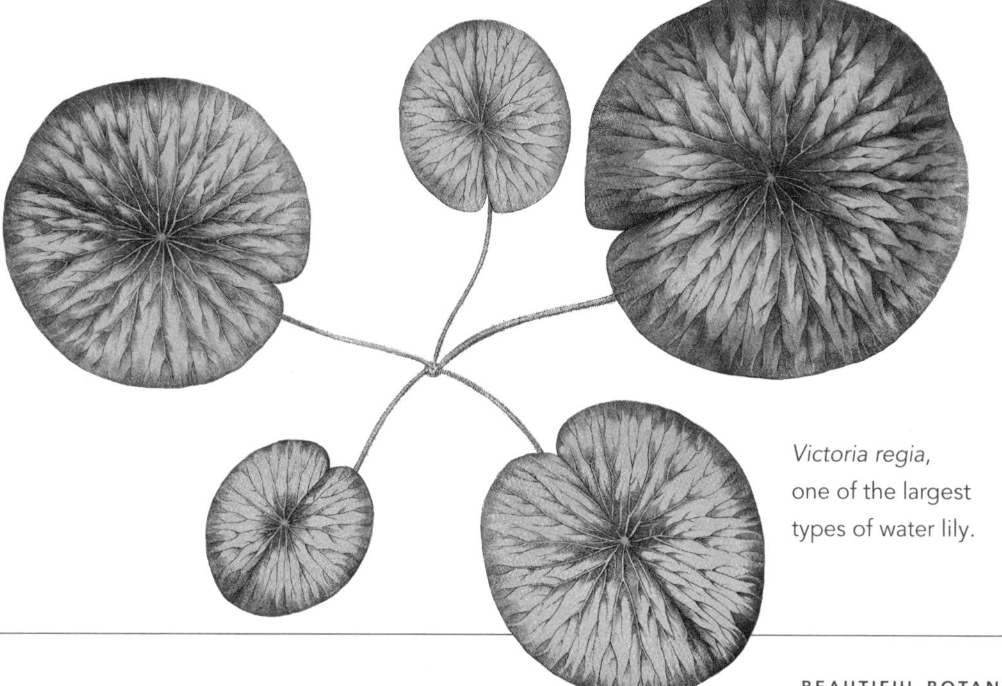

Victoria regia, one of the largest types of water lily.

Whodunnit? Poisonous plants

Don't touch!

The gympie gympie, or stinging tree (*Dendrocnide moroides*), grows in the coastal Queensland rainforest. The shrub, which typically grows to 3m (10ft), is covered in small hairs that are sharp, hollow and loaded with irritant chemicals. When a hair is touched, the tip breaks off, injecting poison that results in severe burning and stinging that can last for days if not months. The pain is excruciating, with one researcher who has been stung calling it comparable to being 'burnt with hot acid and electrocuted at the same time'. There is no known antidote (although you can reduce the effects by trying to remove the hairs from your skin). The shrub itself looks deceptively ordinary, with large heart-shaped leaves and pink mulberry-like berries. These are edible but, as they are also covered in hairs, it is no surprise that they aren't popular.

A beautiful killer

Belladonna, also known as deadly nightshade (*Atropa belladonna*), has been used over the centuries as a medicine, a cosmetic and a poison. Famed botanist Carl Linnaeus had its measure when he named it in his *Species Plantarum* in 1753. Atropa was the eldest of the three Fates in Greek mythology and was the one who decided when mortals were to die. *Belladonna* means 'beautiful woman' in Italian, which refers to the story that women during the Renaissance used the juice from the berries in eyedrops to dilate their pupils, in order to look more seductive. By the way, belladonna is also said to help witches fly.

Atropa belladonna

Wolf Man

'Even a man who is pure in heart,
And says his prayers by night,
May become a wolf when the wolfbane blooms,
And the moon is full and bright.'

This was written by Curt Siodmak, horror film director and creator of the Wolf Man legend. Wolf's bane is another name for monkshood (*Aconitum*). It was used to poison wolves in the 18th century, and then became intertwined with werewolf lore. Wolf's bane is said to be able to trigger someone into becoming a werewolf (during a full moon, as above) or otherwise could be used to weaken them. In the Middle Ages, it was prescribed to patients who suffered from lycanthropy, which is the delusion of being a wolf, with sometimes fatal results.

Aconitum napellus

A bed of murderers

The murder weapons in famed crime novelist Agatha Christie's many mysteries can often be traced to that most bucolic of places – the garden. She was a keen gardener and knew her plant poisons, even those that weren't immediately obvious. After all, she trained during the First World War as an apothecary assistant. There's cyanide, produced in the fruit stones of peaches, nectarines and other members of the *Prunus* genus. There's yew, the bitter taste of which was concealed in that most unlikely of places, a jar of marmalade, in *A Pocket Full of Rye*. Digitalin, a poison found in foxglove, was the lethal weapon in four books. 'The active principle in *digitalis* may destroy life and leave no appreciable sign,' notes Dr Gérard in *Appointment with Death*. Others include a lethal dose of coniine (from hemlock, *Conium maculatum*), extracts from yellow jasmine (*Gelsemium sempervirens*), and aconite from the roots of monkshood (*Aconitum*). The garden suddenly looks very guilty indeed.

Other poisonous pretties

- Lily of the valley (*Convallaria majalis*)
- Lupin
- Daffodil (*Narcissus* spp.)
- Laburnum
- Oleander (*Nerium oleander*)
- Castor oil plant (*Ricinus communis*)

Forbidden fruits (and plants)

Anti-Antipodean

Cacti
More than two dozen varieties, including prickly pear, are prohibited as invasive plants in some states, including Queensland, New South Wales and Victoria. Cacti are not native to Australia.

Banana passion fruit
Passiflora tripartita
This aggressive vine with pink tubular flowers is considered a noxious weed in some parts of the world where it is not native.

Hell no!

Tree of heaven
Ailanthus altissima
Sometimes called 'Tree of hell' instead, this Chinese native is banned in the EU, the UK and NI (and listed as unwanted by at least 30 states in the USA) because of its highly invasive nature.

Banned in the USA

Yellow iris
Iris pseudacorus
What has been called the 'unfortunately attractive' (and vigorous) plant is prohibited or listed as a noxious weed or quarantine species in Connecticut, Massachusetts, New Hampshire, Montana, Oregon and Washington.

Olive tree and mulberry tree
Olea europaea and *Morus alba*
Both fruit-bearing olive trees and fruitless mulberries have been banned in Las Vegas since 1991 because their copious amounts of pollen cause severe allergy symptoms.

Wild privet
Ligustrum vulgare
The classic hedging plant is seen as an invasive pest by many states, including Maine, where it is on the Do Not Sell plant list, and Tennessee, where it is illegal to sell.

Opuntia spp.
Ailanthus altissima
Ligustrum vulgare
Olea europaea

Tussie-mussies

Floriography: The language of flowers

There has long been a belief that certain plants encapsulated a certain feeling or sentiment – the ancient Greeks, for instance, wore laurel crowns as a symbol of victory – but it was the Victorians who turned this into something close to an obsession. The 'secret flower language', known as floriography, became a horticultural code for communicating various sentiments and emotions. Small bouquets were used to transmit this petal Morse code and were known as tussie-mussies: 'tussie' being a nosegay, or posy, and 'mussie' referring to the moss wrapped around the stems to keep them fresh. The original 'language', which appeared in books in France and England throughout the 1800s, has been elaborated over the years, not least by florists eager to sell bouquets for any and all occasions.

Hidden meanings

The first thing to say is that one flower can mean many things depending on which source you consult. For instance, the meaning of a hydrangea has been seen as mostly negative, including boastfulness, false pride, frigidity and heartlessness, but it is also listed as communicating gratefulness and appreciation. Meanings seem to revolve in some cases around appearance, as in a cabbage which looks (sort of) like a bundle of cash and so it means 'profit'; or a walnut with a nut that resembles the human brain and therefore stands for 'intellect'. But meanings also draw on folklore, myth, medicinal and other properties. Daffodils, with their Latin name *Narcissus*, were never going to do well and stand for 'egotism'. It is recorded that pennyroyal, rue and tansies were used by women who were trying to induce a miscarriage and their meanings are listed as 'you must leave', 'disdain' and 'I declare war against you'. Then there are some mysteries, such as the meaning of pineapples, which had long been seen as a symbol of hospitality but in some lists were re-interpreted to mean 'you are perfect'.

Say it with a posy

- **Daffodil** (*Narcissus*): Self-love
- **Scarlet geranium** (*Pelargonium*): Stupidity
- **Marigold** (*Tagetes*): Grief, despair

BOUQUET MEANING:
'Your self-love and stupidity exact my pity'

- **Daisy** (*Bellis perennis*): Innocence
- **Wallflower** (*Erysimum*): Fidelity in misfortune
- **Tulip** (*Tulipa*): Declaration of love

BOUQUET MEANING:
'Your innocence and fidelity in misfortune have caused me to declare my love for you'

- **Rose bud** (*Rosa*): A young girl
- **Yellow lily** (*Lilium*): Majesty
- **Lilac** (*Syringa*): First emotions of love

BOUQUET MEANING:
'I confess your majestic beauty has awakened my first emotion of love'

A rose is a rose

The three types

Wild roses
Also called 'species' roses, wild roses were first cultivated in China some 5,000 years ago. Most have five petals, bloom once a year and are fragrant, with thorny stems and hips (if not deadheaded). Types include: dog rose (*Rosa canina*), field rose (*R. arvensis*), Moyes rose (*R. moyesii*), sweet briar (*R. rubiginosa*), Carolina rose (*R. carolina*), swamp rose (*R. palustris*), barberry-leaved rose (*R. persica*).

Rosa carolina subsp. *carolina*

Rosa campanulata alba

Old garden roses

Also called heirloom, antique or historic roses. This class of roses has been around since before 1867, which is when *Rosa* 'La France', considered to be the first hybrid tea, was introduced (see page 32). Types include: Gallic rose (*R. gallica*), Damask rose (*R.* × *damascena*), cabbage rose (*R.* × *centifolia*), hybrid perpetual, hybrid tea.

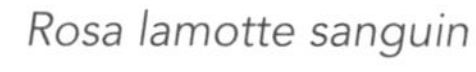

Rosa lamotte sanguin

Modern garden roses

Many modern roses have old garden ones as their ancestors but were created after 1867. Moderns have a wide range of flowering characteristics, including shrub, double flowers in clusters, ramblers and ground cover. Two main groups include hybrid tea (the 'long stem' rose sold widely today) and floribunda (smaller flowers, in large clusters of ten or more).

The oldest rose

The earliest rose species to be described for science was a wild rose fossil found in the Florissant Fossil Beds in Colorado. This is a rich fossil bed, now a US National Monument, in an area where volcanic eruptions occurred some 34 million years ago. The Hayden US Geological Survey and Princeton Scientific Expedition visited the area in 1877 and collected fossils. One of these, which shows a rose stem and leaves, was named by pioneering paleobotanic Charles Léo Lesquereux as *Rosa hilliae*. The name honours Charlotte Hill, a pioneer homesteader, naturalist and amateur fossil collector who lived at Petrified Stump Ranch near the beds. She built up a small museum and hosted many scientists, sometimes giving away part of her collection.

The modern rose

The 'La France' rose, when it was introduced in 1867 by French rose hybridizer Jean-Baptiste André Guillot (1827–93), made history in the rose world. From that moment, all roses that existed before were classified as either wild or 'old garden' roses. All roses introduced after were 'modern'. This remains the same today, so your 'modern' rose may not be that modern at all. What made 'La France' special is that it is seen as the first hybrid tea rose, with its urn-shaped, high-centred flowers. Its parentage is a subject for debate – with many claiming that Guillot himself wasn't entirely sure – but it is often said to be the result of two madames – a hybrid perpetual named 'Madame Victor Verdier' and a tea rose, 'Madame Bravy'. Another possible parent mentioned was 'Madame Falcot'. Whatever the truth, it is an exceptionally lovely rose, with elegant silvery pink petals and a strong, intense perfume.

Rosa 'La France'

Rose water, to drink or spray

- Use fragrant petals from homegrown roses (no pesticides or dyes used).
- Rinse and place them in a pot.
- Add enough water so they are just covered.
- Simmer, then turn down so just below simmering (don't bring to the boil).
- Continue until the petals lose most of their colour.
- Remove from the heat and strain the petals, giving them a squeeze.
- Pour into a glass jar or a spray bottle and store in the fridge.

With love

Why are roses, of all flowers, associated with undying love? It's probably down to Aphrodite, the Greek goddess of love and beauty. She adored her mortal lover and all-out hunk Adonis, and the legend goes that when she discovered a murder plot against him, she ran to warn him, cutting her ankles on a rose bush in the process. Her blood turned the white petals red. Thus a red rose has long been a sign of love and devotion. Sadly, she was too late to save him, as he had been gored by a wild boar.

The China rose

Roses have been cultivated in China for millennia, with the ancient rose gardens being only of the wild varieties. It is not known when, exactly, the China rose (*Rosa chinensis*) appeared in gardens, but it was first seen in art in the 10th century. It was introduced to Europe at the end of the 18th century, as were so many other flowers from China. China roses were heartily embraced as they were repeat bloomers, which, until then, was a rare feature in roses. They also had an extraordinary colour trick, often darkening with age. Most roses before then only faded away; now there was a whole new palette.

'Of all flowers, methinks a rose is the best'

William Shakespeare and John Fletcher, *The Two Noble Kinsmen* (1634)

World of bulbs

Category 'bulb' – who's in?

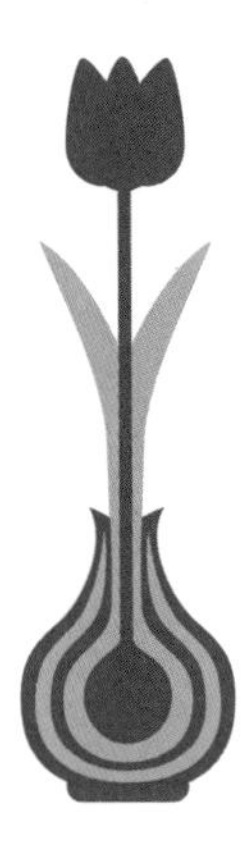

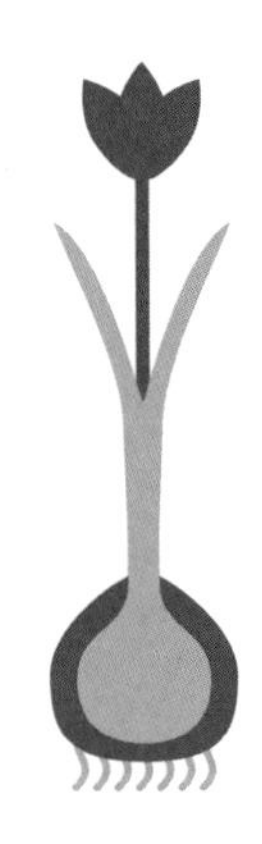

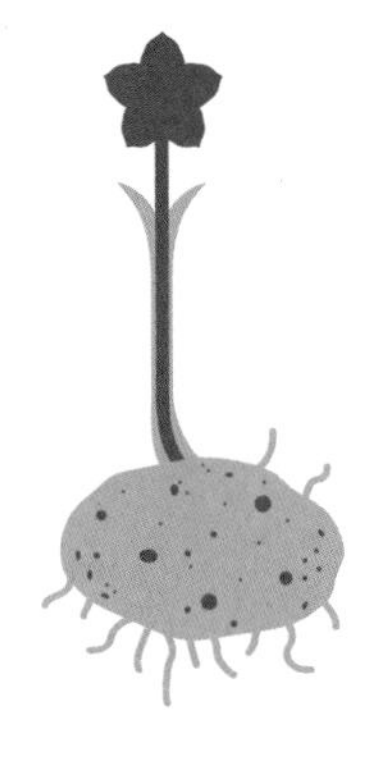

A true bulb has a series of rings, called scales, which are modified leaves that are used to store food during dormant periods. They have a 'basal plate' and often have papery skins. They form a new layer every year. Examples: alliums, garlic, daffodils, tulips.

A corm looks similar to a bulb but does not have rings, or scales, and instead is made up of one piece of stem tissue that stores nutrients. Each flowering depletes the corm and a new one forms on top of the old one as a replacement. Examples: crocus, freesia, grape hyacinths, crocosmia, bananas.

A rhizome is a swollen underground stem that stores nutrients. It grows horizontally near the soil surface. Examples: cannas and many irises.

A tuber is a swollen stem or root that stores food. Shoots develop from the buds, or 'eyes', and several plants can be obtained from one tuber. Examples: dahlias, cyclamen, tuberous begonias and potatoes, which are stem tubers.

The tulip that conquered all

Its name is *Tulipa* 'Semper Augustus', aka the King of Tulips, and its regal petals are milky white with remarkable marbled veins of crimson. It is considered to be the most famous tulip in history, and when you consider that this history includes the crazed bulb-buying of *Tulpenwoerde*, or tulip madness, in The Netherlands in the early 1600s, that is saying something indeed. Its story, as well as the wider history, is a morality tale of glamour as well as greed. This was the era when the newly enriched Dutch (this was their golden age) had an insatiable desire to own tulips to grow and display, the more unusual the better. The rarest and most striking tulips of all were the 'broken tulips', with their flame-patterned petals. No one really understood what was behind the 'breaks', and growers tried all sorts of things – including adding paints to soil – to create them.

It is not known who created the first 'Semper Augustus' but its rare beauty was undeniable, and its star quality was seen in catalogues as well as still-life flower paintings that were all the rage. Amsterdam's Tulip Museum notes that in 1633 one bulb was said to sell for 5,500 guilders (more than three times the annual earnings of a merchant) and in 1636–37, it was said that 10,000 guilders were offered (enough to buy a substantial home in the city). Of course, the bubble burst, but in the case of 'Semper Augustus' there was still another twist to the tale. We now know that 'tulip breaking virus' was the cause of its beautiful petals: the disease creates gorgeous patterns but also weakens the bulb and all offspring until, at some point, it stops flowering. Thus, 'Semper Augustus' no longer exists: what made it special is what killed it. Growers have tried to recreate its beauty, but so far to little avail. The prestige and high prices may be long gone, but that flower, painted all those years ago, still has the magic.

Tulipa
'Semper Augustus'

Good fortune for a long-distance traveller

The bulb known as the Sacred Chinese Lily is neither Chinese nor a lily. Instead it's a daffodil, specifically *Narcissus tazetta orientalis*, small with scented white petals and a golden yellow cup, beloved in China and an essential part of its lunar new year festivities. This little bulb has quite a travel tale to tell. It's not native to China but hails from the Mediterranean and western Asia, and has warm-weather tastes. It has been seen depicted in an ancient Egyptian tomb and many believe it might have been the Rose-of-Sharon in the Bible's Song of Solomon. No one knows how it got to China: one theory is by boat, another more romantic version is that it arrived by camel caravan along the 6,500-km (4,000-mile) Silk Road. Whatever its route, it was already being grown on a large scale by the start of the Song Dynasty (960 CE). The bulbs, which flower early enough to match the variable new year timeframe, have only become more popular in China. They even travelled with Chinese workers during the Gold Rush of the 1850s, and so found their way to the pioneer American West, where they can still be seen blooming. The bulbs, which can be grown in a shallow tub with pebbles, are tended carefully to get them to bloom exactly on Chinese New Year. If successful, then it is said to bring extra wealth and good fortune all year round.

N. tazetta orientalis

Naturalizing bulbs

1. Scatter bulbs randomly over the area and plant them where they fall, though make sure they are not too close to one another.

2. Dig planting holes with a trowel or bulb planter. Make the holes three times the depth of the bulb.

3. Break up the soil from the turf removed with the trowel/planter and use this to backfill after the bulb is in the hole.

4. Replace the turf so it is even with the lawn surface.

5. Wait for the seasons to change.

Bulb lasagne

The idea behind what is called a bulb lasagne is to create a container that has a succession of different blooms for weeks on end. The larger the bulb, the deeper it is planted, and you need to create layers in your pot accordingly. Flowers could start out in early spring with snowdrops (*Galanthus* spp.), then daffodils, then tulips, then alliums. As many varieties of the same bulb type, such as daffodils or tulips, flower at different times, you could create a daffodil or tulip lasagne.

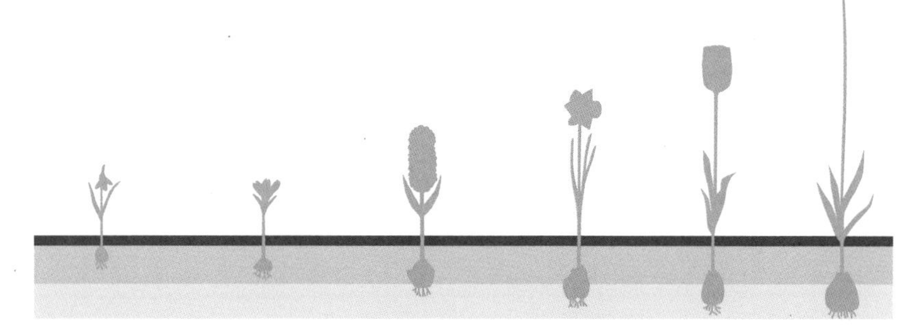

Botanical illustrations

An eye for detail

Botanical illustration is a marriage between science and art, and, before photography, was the only way to depict the anatomy of plants. The drawings and paintings, drawn from herbarium specimens or live plants, were much more than horticultural portraits. The more scientifically oriented depict an entire life cycle as well as natural habitats and some dissections. Many are in black and white, for economy, but also to emphasize intricate details that colour may have distracted. At the more artistic end are watercolour paintings that are still scientifically correct but emphasize the aesthetic value of a flower or plant without all life-cycle information.

The 'Raphael of Flowers'

Pierre-Joseph Redouté is the most famous botanical artist of them all. Born in what is now Belgium in 1759, he set off at age 13 to explore the world, visiting museums and working as a jobbing artist. This is when he fell in love with painting flowers. In 1782 he moved to Paris to join his brother who was a stage set designer. He became a regular at the Jardin des Plantes (then called the Jardin du Roi), where he was soon illustrating prestigious botanical publications.

Politically, he had the ability to shape-shift. He was art tutor to Marie Antoinette and draughtsman and painter to the Queen's Cabinet. After the Revolution, he became a favourite of Empress Josephine Bonaparte, who appointed him to paint flowers.

Memoirist François-Joseph Grille describes him during his heyday: 'A dumpy body, limbs like an elephant's, a head as heavy and flat as a Dutch cheese, thick lips, a hollow voice, crooked fingers, a repulsive look, and beneath the skin an extremely delicate sense of tact, exquisite taste, a deep feeling for art, a fine sensibility, nobility of character, and the perseverance needed for the development of genius: such was Redouté, who had all the pretty women in Paris as his pupils.'

Redouté produced over 2,000 paintings depicting more than 1,800 species. He was heralded as the 'Raphael of Flowers' and by 1819 had a painting in the Louvre. However, his last 20 years were impoverished. He died in 1840 at the age of 80, and his work has never gone out of fashion.

Rosa × damascena

Ah sunflower!

Devoted follower

The Greek myth associated with the sunflower is often called a love story, though it is a rather one-sided one. The water nymph Clytie fell madly in love with Helios, the sun god (*Helianthus* is the Latin name for the sunflower), spending her time gazing up at him adoringly. He went along with this for a while but soon fell in love with another nymph. Clytie displayed extraordinarily poor judgement by taking revenge, telling the nymph's father about it. He promptly buried his daughter alive. Helios took his revenge by turning Clytie into a flower who showed her continued devotion by looking up at him, her gaze following him throughout the day. The myth has many versions. In the original account Clytie was turned into a heliotrope, but in later retellings (and in the best-known version today) the flower became a sunflower. And it is the case that young sunflowers do look to the sun all day long.

These incredible flowers...

- were domesticated more than 5,000 years ago by Indigenous Americans.
- were used by Incas and Aztecs as a symbol of the sun gods.
- were used in clean-up operations at both Chernobyl and Fukushima, to remove toxic elements from the soil.
- can be used to produce biofuel by mixing their oil with diesel.
- are used to produce sunflower oil, with more than 50 per cent of worldwide production coming from Ukraine and Russia. The war in Ukraine caused a shortage.

Harvest and roast

Harvest the seeds by leaving the flower head on the stem for at least two weeks. Once the petals have faded, cut the heads off and store them in a warm and dry place for another week. Then place on a large piece of paper and rub the head: the seeds should fall easily on the paper.

If you want your roasted seeds to be salty, bring the in-shell seeds to the boil in salted water, then turn down to a simmer for 15 minutes or so. Spread out on a tray in a single layer and roast in an oven preheated to between 180 and 200°C (356 and 392°F) for 10–20 minutes.

Helianthus annuus

Strange and wonderful plants

A parasite with a big difference

The world's largest flower, *Rafflesia arnoldii*, or corpse flower, is a show-stopper, at least of a sort. Its blooms are 60–90cm (2–3ft) across and carry a smell akin to rotting flesh. The five petals also look a little like meat gone bad – orangey-red with white-ish dots. It's parasitic, hiding for most of its life inside the chestnut vine (*Tetrastigma voinierianum*) in the rainforests of Borneo and Sumatra, taking on its host's water and nutrients. The blood-red bud emerges outside the vine, taking almost a year to grow, before bursting open in all its smelly glory for only a few days.

Rafflesia arnoldii

The plant that will not die

At first glance, the low-growing plant *Welwitschia mirabilis*, found only in the Namib desert on the west coast of southern Africa looks to be dying, its green leaves surrounded by a carpet of brown ripped ones. But it is very much alive and will have been so for on average 500–600 years (carbon dated), though some plants are believed to be 2,000 years old. Thought to be a relic from the Jurassic period, it is a gymnosperm with two permanent leaves, broad and strap shaped, that are never shed but continually keep on splitting. It survives by absorbing water from sea fog and via deep groundwater. It is known in Angola by the name *n'tumbo* and in Afrikaans as *tweeblaarkanniedood*, or 'two leaves that cannot die'. The seed cones can be eaten raw, baked or roasted, while wildlife, including oryx and rhino, chew the leaves for water.

No publicity please

Lithops, also called 'living stones', are reluctant celebrities. They survive in the driest areas of South Africa and Namibia by keeping a low profile (indeed, often partially buried) and staying small. Each plant is made up of two succulent leaves, fused in the shape of inverted cones that look like pebbles. The leaves store enough water to survive for months without rain, and the tops of the leaves are partially translucent, offering a 'window' for photosynthesis to occur. The flower is daisy-like and pops up once a year between the two 'stones', meaning that, for a short time, the plant is very visible.

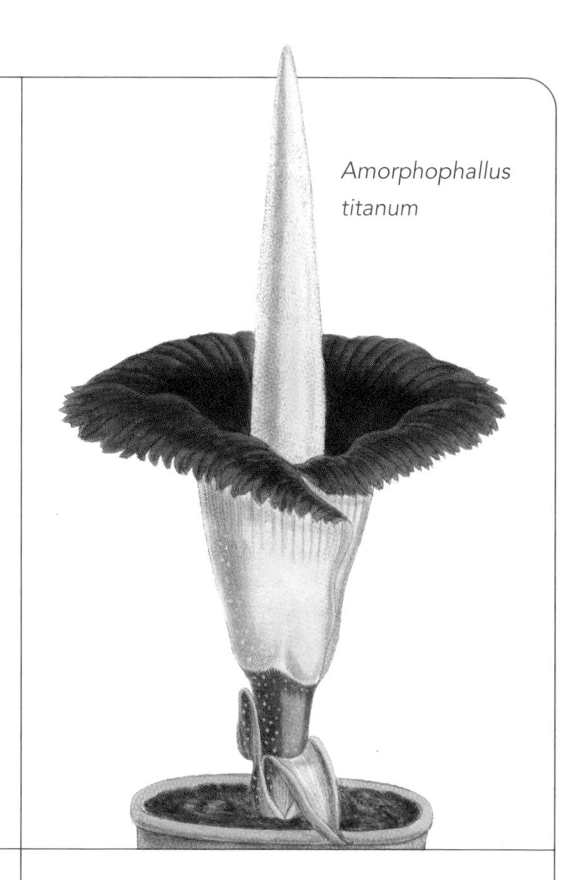

Amorphophallus titanum

Mmmmm. . .what's that smell?

- *Cercidiphyllum japonicum*, or Katsura tree, smells like toffee apples and caramel in the autumn.
- *Helichrysum italicum*, or curry plant, smells just like curry powder.
- *Agastache foeniculum*, or anise hyssop, smells like liquorice.
- *Azara microphylla*, or box-leaf azara, is a small evergreen tree from Chile with flowers smelling of vanilla.

Ughh. . .what's that smell?

- *Amorphophallus titanum* (which means 'giant misshapen penis'), aka the titan arum, smells like rotting flesh.
- *Pyrus calleryana* 'Bradford', a pear tree with flowers described as smelling like dead fish.
- Female *Ginkgo biloba* tree, produces a smell in the autumn that has been likened to rancid butter or vomit.
- *Lysichiton americanus*, from America's Pacific Northwest, is known as 'yellow skunk cabbage' for a good reason.

Pyrus × serrulata

Chapter 2

Curious Horticulture

'Curiouser and curiouser!' cried Lewis Carroll's Alice in Wonderland. Many of us feel this way about our own garden wonderlands. You'll find plenty to sate your horticultural curiosity in this chapter, with its mix of the practical and fantastical, historical and inspirational.

Garden style

Formal

A formal garden style uses structure involving mostly straight lines or linear geometric shapes. Formal gardens are known for their symmetry, hedging and muted colour palette. They aspire to look controlled, tidy and managed, creating a sense of soothing calm.

Roman

- Courtyard garden, or vivarium, with geometric layout.
- Low hedging, often flowering, plus often roses, citrus, topiary.
- Features include statues and water features, including a central pool.

Persian/Islamic

- Divided into four equal sectors (representing sky, earth, water and plants) by water channels. While this sounds simple, the gardens tend to be intricate.
- Adapted for structure of Islamic *charbagh* (quadrilateral garden) using water/mosaic/tiling/ornamentation.
- A *chadar* water feature, with water chutes that link pools.

Mughal

- Influenced by Persian and Islamic gardens.
- Divided into four parts, separated by water channels.
- Places of rest and reflection, a reminder of paradise.

Monastery

- Medieval, geometric layout, with beds enclosed in wattle fences or hedges.
- Physic gardens with medicinal herbs growing inside regimented beds.
- Also featured flowery meads, edible herbs and trellises.

Knot

- Elizabethan, with low hedging in an intricate curving pattern to resemble knots.
- Symmetrical and orderly in structure.
- Beds filled with perennial flowers or coloured gravel.
- Many adaptations, including French *parterre de broderie*, and used in small modern spaces.

Baroque

- Championed by Louis XIV in France and landscape architect André Le Nôtre.
- Extravagant and geometric, creating long avenues and perspective, with water in various forms.

The Parterre at the Palace of Versailles, France

Informal

Informal gardens feature curved lines for borders and paths, to create flow, with no central axis. They often seek to imitate natural landscapes, evoking relaxation and a connection to nature.

Chinese

- Idealized natural style that has evolved over thousands of years.
- Winding paths and bridges over ponds.
- Ponds, rock works, trees and flowers, as well as various buildings.

Japanese

- Designed to highlight the natural landscape.
- Mosses, gravel, rocks, water and harmonious planting.
- Often designed to be seen from specific views.

English Landscape

- 18th-century movement to get away from formality.
- Ambitious taming of nature with picturesque landscapes.
- Sweeping grassy vistas, rivers and trees.
- Best-known proponent is English landscape gardener Capability Brown (see page 180).

Cottage garden

- Small gardens attached to homes of workers.
- Mix of herbs, fruit trees, ornamental flowers.
- Seen as English and French, with Monet's Giverny garden a classic example.

Meadow/Prairie/Dry/Native

- Seeks to emulate nature in a domestic or controlled space.
- Flowing and natural, aim is to look unplanned.
- Native plants offer an opportunity to 'go back to the past', and to be more environmentally friendly.

New Perennial

- Features repeating drifts of perennials and grasses.
- Designed to look good in all seasons, leaving seedheads on in winter.
- Best-known designer is Dutch plantsman Piet Oudolf.

Two Girls Enjoying the Evening Cool in a Garden (c. 1765) by Suzuki Harunobu

Mix of formal and informal

Arts & Crafts

- Influential 1930s English movement epitomized by gardener Gertrude Jekyll (see page 170).
- Tends to feature geometric structure and local materials.
- Informal planting style that emphasizes movement and colour.

Minimalist/Modernist

- Stripped-back style emphasizing simplicity and order.
- Geometric forms, often featuring gravel and/or rocks.

More notable styles

Pelargonsjuka: Scandi style that involves using pelargoniums in pots, indoors and outdoors.

Xeriscaping: Climate-change and desert inspired, planting that needs no watering.

Outdoor kitchen: Blending the boundaries of inside/outside, Australian style.

Indoor garden: Indian 'plant influencers' want grey urban homes to become green inside.

Eco-gardening: Designing a garden for wildlife, be it birds, bees or toads.

Water Lilies and Japanese Bridge (1899) by Claude Monet

'My garden is my most beautiful masterpiece.'

French painter Claude Monet (1840–1926)

The great lawn obsession

Defining lawn

The word 'lawn' – meaning an area of closely cropped grass – dates from the 16th century, but there is dispute about its origin. Some say it is from the old French word *launde* or *lande*, or Middle English *lande*, for a place for wildflowers; or the Breton word *lann*, for heath.

The grass beneath our feet

Formal and fine lawns: A blend of finer grasses to create a soft, thick carpet. Seeded from chewing's fescue, red fescue and bents. Slow growing and can be mown at a low height.

General purpose: A mix of hard-wearing grasses that can handle regular traffic. A blend of dwarf perennial ryegrass, creeping red fescue and bents. Fast growing and needs regular mowing.

Shady: For a light to medium shade; a mix of rye, fescues and bents. Not hard wearing.

Great grass alternatives

Chamomile: The preferred variety is the non-flowering Roman chamomile *Chamaemelum* 'Treneague', which is low growing and creeping. Forms a dense mat of scented foliage.

Moss: Inspired by Japanese gardens, a moss lawn needs a shady, moist spot. Verdant and beautiful.

Clover: Lawns of white clover (*Trifolium repens*) are seen as a more sustainable alternative to traditional grass. Drought-tolerant, good for biodiversity and low maintenance.

Flower meadow: Use annuals (one year only) or a perennial mix of wildflowers. Requires low-fertility soil. Yellow rattle (*Rhinanthus minor*) is a parasite meadow flower that weakens grass and can be planted as part of the scheme if planning a mixed flower/grass lawn.

Mowing patterns

The key to creating perfect lawn stripes, or other patterns, lies in the mantra: mow high, mow dry, mow sharp. The height of the cut is central, as shorter grass bends less and so reflects less light (and therefore looks darker). For a more visible pattern, cut grass to 7.5–8cm (3–3½in) instead of the usual 5–6cm (2–2½in).To make the pattern even more obvious, use a lawn roller, which can be done separately or with a mower with a built-in roller.

For wildlife, you need to avoid mowing in summer but should mow in autumn to prevent tussocks and unevenness. As grass grows slowly in winter, very occasional (or no) mowing suffices. Movements such as 'No Mow May' advocate leaving your lawn unshorn throughout the month (and beyond), or at least mowing less and leaving long grass in places.

Chequerboard

Curves

Stripes

Concentric circles

How the lawn spread across the world

1400s: Wealthy families in England enclose gardens for security, usually with a square central lawn made of meadow turf, to create a 'flowery mead'.

1600s: Louis XIII introduces the *tapis vert* at Versailles, France, which, under Louis XIV and landscape designer André Le Nôtre, is expanded to become the 'great lawn' or 'Royal walk', 335m (1,000ft) long and 40m (130ft) wide.

1700s: A lawn becomes an essential element of English and French garden design and a status symbol for the wealthy, who employ workers to scythe the grass.

1730s–90s: The first version of a French formal garden in China was created in the Old Summer Palace of Qianlong in Beijing, also known as the 'Versailles of the East'.

1806: US President Thomas Jefferson, a keen horticulturist, kicks off what would become a lasting American obsession by introducing a European-style lawn at his Monticello estate in Virginia.

1800s: As the British empire expands, colonists introduce sweeping lawns in places such as India, South Africa and Australia.

1830: Edwin Budding, an engineer from Thrupp in Gloucestershire, UK, adapts a carpet cutter to create the first lawnmower.

1839: Lawns become more common in China after the First Opium War, when ports such as Shanghai and Guangzhou were opened to foreign countries, including Britain and the USA.

1870s–80s: The US government backs 'lawns for all' with a display at the 1876 Philadelphia World Fair. American magazines begin to carry articles on lawn growing and maintenance.

1893: James Sumner of Leyland, Lancashire, UK, designs the first steam-powered lawnmower.

1902: The firm Ransomes in England produces a petrol mower commercially.

1938: The American love affair with the lawn is enhanced by the Fair Labor Standards Act, which institutes the 40-hour work week. Many use newly freed-up weekends to start, tend to and enjoy lawns.

1945: The pesticide DDT, or dichlorodiphenyltrichloroethane, goes on sale in the USA for a variety of uses, including lawns. It is banned there in 1972 and in the UK in 1986, for its negative impact on animal and human health.

1949: The newly created People's Republic of China turns some colonial lawns into community assets for local people. Its General Greening campaign urges that land designated for pleasure is used for more utilitarian purposes such as fishery or agriculture.

1950s: The rotary power mower surges in popularity in the USA, making it easier for the growing middle classes living in the burgeoning suburbs to obsess over lawns.

1964: Flymo introduces the hover mower in the UK.

2000s: Climate change is the catalyst for reinventing the lawn. Sustainable alternatives that require less water and resources, such as meadows and clover lawns, become more popular.

2005: Robotic lawnmowers are released onto the market.

Sow from seed

1. Clear the ground then dig over to remove perennial weeds. If the ground is compacted, dig over to 20cm (8in) in depth.
2. If the soil is light and free draining, add peat-free organic compost or well-rotted manure.
3. Leave for at least two weeks and remove any regrowth.
4. Firm the area by walking on it with small shuffling steps or using a roller.
5. Rake the area in all directions to level.
6. Add an organic plant-based root feed if needed.
7. Measure out seeds with the quantity depending on the size of area (it will say on the seed packet).
8. Scatter half while walking up and down in one direction, and the other half walking up and down in the other direction, or use a spreader. Lightly rake to mix seeds with soil.
9. Water with a light spray (if it's not raining). If practical, cover to protect from birds.
10. Water every few days, as necessary.

Herb garden

Layout: a trio of triangles

Consider planting a herb (and spice) garden in three triangular sections, separated by theme:

Triangle 1: Culinary

Annual: Basil, coriander, garlic, summer savory, mustard, borage, dill, rocket, nasturtium, marigold, red mountain spinach.
Biennial: Angelica, chervil
Perennial: Bay, thyme, chives, winter savory, rosemary, sage, parsley, mint, caraway, horseradish, fennel, chicory, lemon balm, lovage.

Triangle 2: Medicinal

Peppermint, ginger, turmeric, chamomile, ginseng, feverfew, evening primrose, echinacea, lavender. Many medicinal herbs have more than one use: peppermint, for instance, aids digestion but also can be used for headaches, nausea, flatulence and some types of anxiety.

Triangle 3: Household

Rue and santolina (anti-moth); rose petals and lavender (in linen bags).
Dyeing herbs: Madder and Lady's bedstraw (red); dyer's coreopsis (orange); woad (blue); weld, goldenrod, great mullein (yellow); woad and weld mixed, yarrow (green); woad and madder (mixed), hollyhock (purple).

The problem with parsley

Parsley can take time to germinate, and that may be the nub of why it has attracted so many superstitions. Indeed, it is seen by some as being positively diabolical. First, it is said a woman of child-bearing age must never buy or plant parsley unless she wants to get pregnant. Virgins who plant it could be expected to be visited by the devil, so they shouldn't get involved. But, even then, there are issues with which day to plant. The safest bet was thought to be for men to plant parsley at 3pm on Good Friday, which is when God is closest (according to Christians at least). If this plan was not followed, the devil would make the parsley roots go all the way down to hell and back nine times before sprouting. That's why it takes so long to germinate. Now you know.

More old wives' tales

- The size of a rosemary bush reflects the power of the woman of the house.
- Put marjoram under your pillow and dream of your future love.
- Basil grows best when there is swearing.
- Sage only grows when planted by a stranger.
- Where thyme grows, the atmosphere is pure.
- To prevent unwanted guests returning, sweep the doorstep and rub with bay leaves.

Keep people at bay with *Laurus nobilis*

Fragrant potpourri

Use rose petals from the scented *Rosa gallica*, or one of the damask roses, as they hold their scent. Just-opened roses usually have a strong scent, while lavender has a strong scent for a significant length of time. Feel free to adapt the recipe below to fit what is in your garden. A herby mixture may include bay leaves, eucalyptus leaves, myrtle, lovage, thyme and/or sage. For a sweeter mixture, try bergamot, violet and/or rose-scented geranium. Gather petals throughout summer and dry in an airy, shady room. Store in separate closed containers.

1. Create the base mixture: In autumn, rub a bowl with oil of cloves and mix in a total of 50g (10 cups) of petals, half of which could be rose, followed by your choice but which could include carnation petals, lemon verbena leaves, lavender, rosemary or marjoram, peppermint leaves, violets.

2. Add fixatives, to prolong the fragrance: Options include orris root (made from rhizomes of irises), vetiver, vanilla beans or pods, oakmoss, angelica root, myrrh gum/resin. For a recipe such as this, you might use a mixture of 30g (1oz) chopped orris root, half a tablespoon powdered orris root, half a tablespoon of gum benzoin, plus a chopped vanilla pod.

3. Enhance the look and scent: Add seedpods, small pine cones, whole cloves, cinnamon sticks, crushed coriander seed, dried citrus, whole or grated nutmeg, bay leaves, chopped ginger, star anise, allspice berries.

4. Mix and cure: Mix it all together in an airtight container. Add a few drops of rosemary or basil oil (or your choice of flower oil), and give it a good shake to distribute the oil and fixatives. Cure by leaving for four to six weeks in a dark place, giving it a shake every few days. Display in a non-metal bowl.

Rosa gallica

Ayurvedic mind/body balance

The traditional Indian system of ayurvedic medicine is all about keeping the mind and body in balance. It promotes wellness via lifestyle, diet and exercise changes. Herbs and spices are an integral part of this.

Ashwagandha (*Withania somnifera*): Small woody plant native to India and North Africa that is helpful in managing stress, improved sleep, muscle growth and male fertility.

Boswellia (*Boswellia serrata*): Also known as Indian frankincense or olibanum. Thought to have anti-inflammatory properties and to reduce joint pain.

Brahmi (*Bacopa monnieri*): A herb believed to be anti-inflammatory, to improve brain function and help symptoms of attention deficit hyperactivity disorder (ADHD). Beware of side effects.

Cumin (*Cuminum cyminum*): Thought to help with digestion and to reduce cholesterol.

Turmeric (*Curcuma longa*): Thought to help heart and brain function, as well as being an anti-inflammatory.

Flower borders

Here are some ideas from what could be much longer lists of plants that are typically found in these types of gardens. Many, many more are available.

Gravel garden

Drought resistant and low maintenance, many gravel favourites are self-seeding.

- *Agapanthus*
- *Hylotelephium*
- *Iris*
- *Nepeta*
- *Salvia*
- *Santolina*
- *Stipa*
- Thyme (*Thymus*)
- *Verbena*
- *Verbascum*
- White lavender (*Lavandula stoechas*)
- *Allium*
- Blue fescue (*Festuca glauca*)

Cottage garden

A kaleidoscope of colour and texture that has become a definitive style.

- Foxglove (*Digitalis*)
- Lavender (*Lavandula*)
- Mock orange (*Philadelphus*)
- *Salvia*
- *Astrantia*
- Cornflower (*Centaurea cyanus*)
- Sweet pea (*Lathyrus odoratus*)
- *Geranium*
- Rose (*Rosa*)
- *Delphinium*

Tropical border

Big leaves, vibrant colours, water-thirsty, with a built-in wow factor. (Note: These are tropical-looking plants for a temperate border, not actual tropical plants – an endless list.)

- Japanese aralia (*Fatsia japonica*)
- *Canna*
- Chinese rice-paper plant (*Tetrapanax papyrifer*)
- Tree fern (*Dicksonia*)
- Jasmine (*Jasminum*)
- Castor oil plant (*Ricinus communis*)

Desert border

Xeriscaping features sculptural, often prickly plants that reduce or eliminate the need for irrigation. For warm, dry climates only.

- *Aloe*
- Fire barrel cactus (*Ferocactus gracilis*)
- Poppy (*Papaver*)
- Brittle bush (*Encelia farinosa*)
- *Agave*
- *Hibiscus*
- Poached egg flower (*Limnanthes douglasii*)
- Desert pea (*Swainsona formosa*)

Lessons in propagation

To propagate or pollinate? That is the question

The two main categories when it comes to propagating plants are 'sexual' and 'asexual'. The former involves pollination and new seeds being created. Asexual propagation is all about using a piece of one plant to create another via cuttings, layering, division and grafting. Often this is easier and faster than pollination, but the new plant will be a clone and so will lack the genetic diversity and hybrid vigour of one grown from seed.

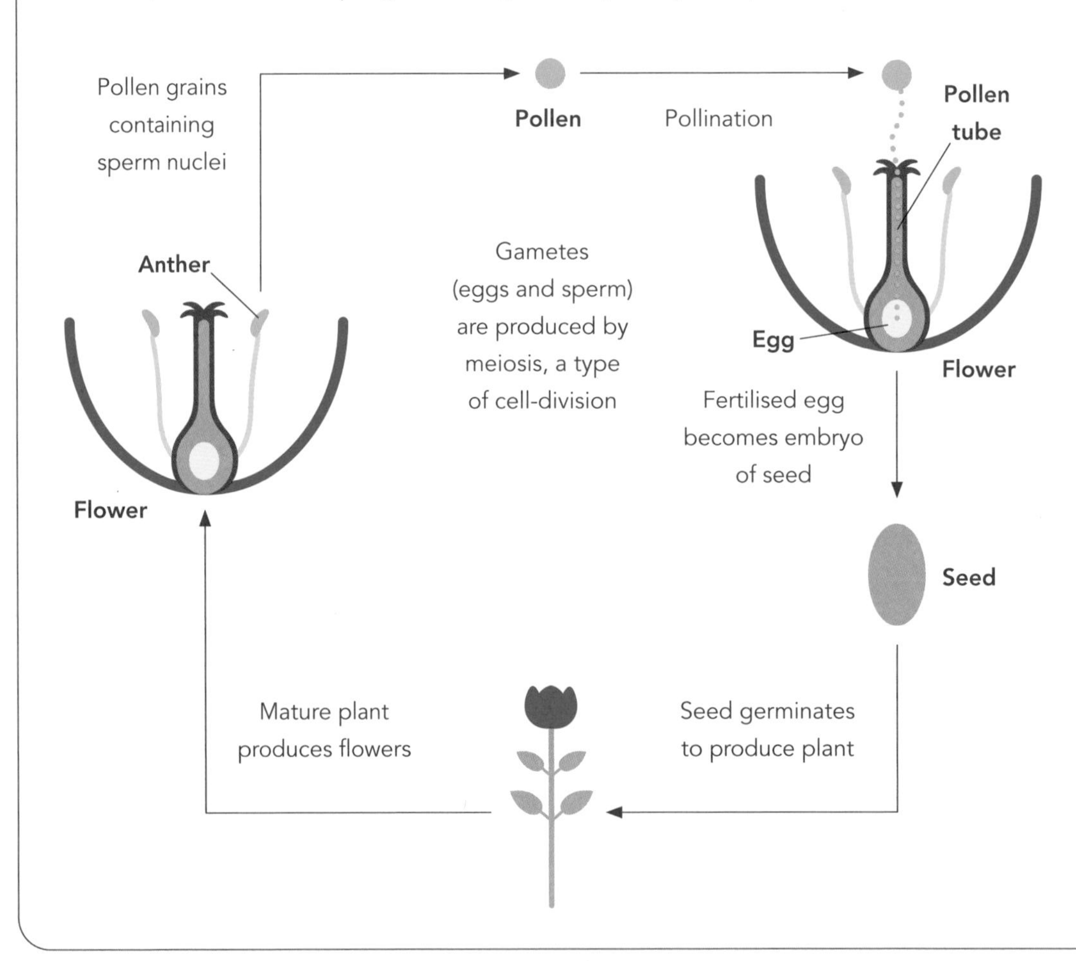

ASEXUAL

New plants are clones

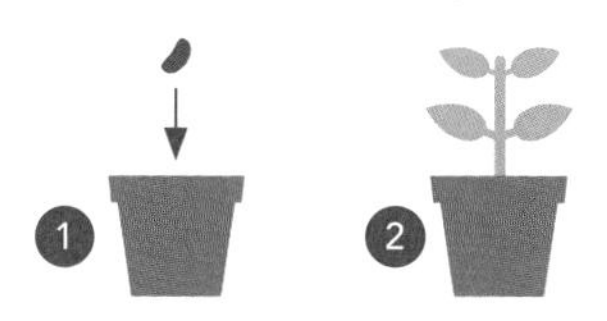

Asexual seeds

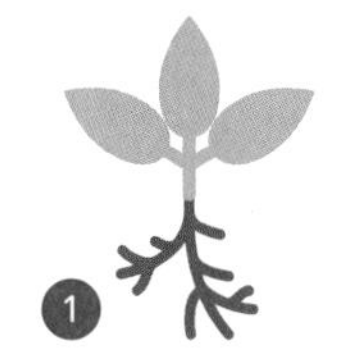

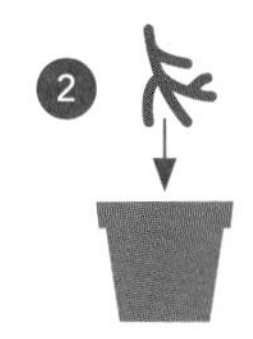

Root cutting

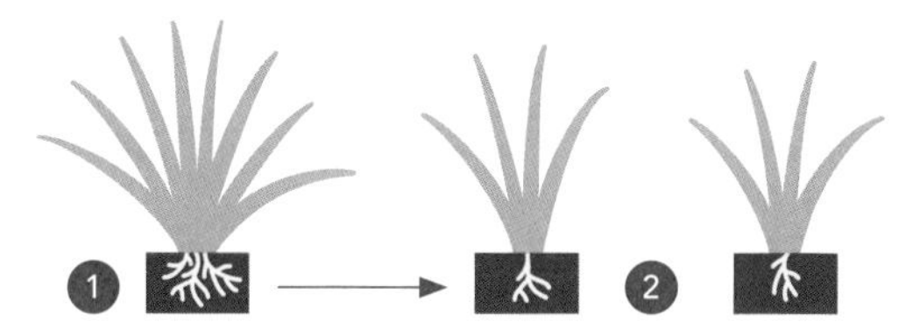

Division

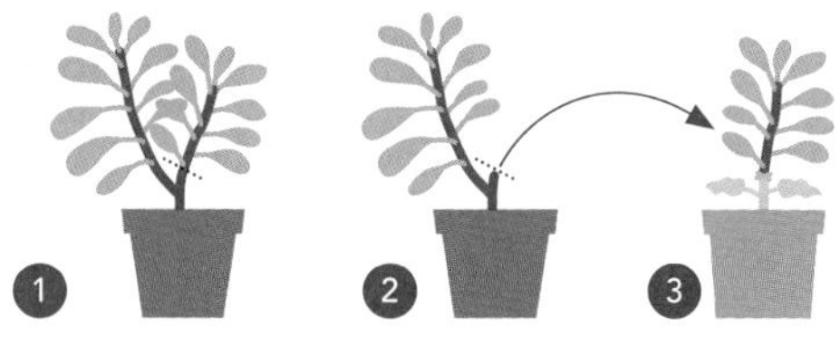

Grafting

Runners

A clone is born

Asexual propagation means, basically, reproduction by a separation of the body into two new bodies. One organism duplicates its deoxyribonucleic acid (DNA) and divides (cytokinesis). And *voilà*, a clone is brought into existence.

Largest vs smallest seeds

The seed of the double coconut palm (*Lodoicea maldivica*), known as the *coco de mer*, is on average 30cm (12in) in diameter and weighs 18kg (40lb). At the other end of the scale, tropical orchid seed is like a speck of dust, weighing in at 4.3 millionths of a gram.

Bounteous birch

Birches can produce an average of 1.66 million pollen grains per catkin, and there can be thousands of catkins on a single mature tree.

Child's play

When is a sandbox not a sandbox?

A sandbox can occupy a child for hours, and it seems that it's one of those childhood places that sticks with us. Adult politicians, business people, artists of all kinds refer to the sandbox as a place for coming up with solutions and learning to get along with others. Canadian artist Robert Genn notes: 'The studio is an extension of the sandbox and the kindergarten playroom… It's a room at the service of a dreamer on her way to becoming a master.' Here's Lebanese-American writer Rabih Alameddine: 'Literature is my sandbox. In it I play, build my forts and castles, spend glorious time.' Even the politicians are at it: 'We are looking for ways where you can have a sandbox, where you have a restricted environment within which people can try new things,' said Singaporean Prime Minister Lee Hsien Loong. It seems a sandbox isn't just a place to play in the corner of the garden: it's an entire life philosophy.

It's a dream

Let's get Shakespeare to do the planting plan for this outdoor theatre:

'I know a bank where the wild thyme blows,
Where oxlips and the nodding violet grows,
Quite over-canopied with luscious woodbine,
With sweet musk-roses and with eglantine.'

This is where, in *A Midsummer Night's Dream*, the faerie queen Titania sometimes sleeps at night. Perfect for a sleepout zone or, even, with a few logs around it, a stage to put on an outdoor play.

Rosa eglanteria var. *luteola*

Salix alba

Watery fun and flow

- Birdbath: fill and watch
- Solar-powered fountain for splashing and sound magic
- Container pond for wildlife watching
- Water wall, created by attaching old plastic bottles or metal cups (tipped) to a trellis, so the water drips from one to another
- Water run: repurpose plastic bottles or recycle those old plastic gutters to create a run for small-boat races.

Make a living willow feature

Create a tunnel, high or low, using living willow. Any willow will do, though *Salix alba* var. *vitellina*, *S. viminalis* and *S. purpurea* are popular. The best time is autumn through to spring (before the willow sprouts leaves).

You will need:

- Living willow whips, width as per your creation but for an arch of at least 2m (6½ft)
- Twine or string
- Spade or garden fork
- Metal pole

1. Mark out the space, removing turf at least in the areas where the willow will be planted.
2. Use the metal pole to create holes in the ground.
3. Create arches at both ends, using two whips on either side, weaving them together at the top.
4. 'Plant' whips every 30cm (12in) or so and about 30cm deep.
5. Use a lattice pattern (every other whip that is planted will lean one way, to be woven through) if you wish. It will be stronger.
6. Tie uprights at the top with twine or string.
7. Stand back, admire and make sure it's well watered.
8. Wait for spring.

Other projects from living willow include backdrops for benches, trellises, forts and dens.

Grow an orchard

What exactly is it?

An orchard is an area of land devoted to the cultivation of fruit trees or a collection of fruit trees. Some orchards are vast commercial enterprises, made up of parallel rows of trees that are grown and harvested as part of farming operations. Domestic orchards have been planted since Roman times. These were called *pomarius*, from the Latin word for 'fruit seller', but it also means 'from the walled garden', and refers to the Roman goddess Pomona, deity of fruit trees and orchards. In Islamic gardens, orchards are called *bustans*. These smaller domestic orchards produce fruit, but they are also are grown for ornamental purposes. There has been a revival in planting community orchards, both for the benefit of wildlife and local people.

Get together and get fruity

The number of traditional orchards is declining sharply (England and Wales have lost almost half of theirs since 1900), which is bad news on several levels. It affects the variety and amount of fruit that is grown and is detrimental to wildlife, as these orchards are great places for bees and other pollinators. It also means fewer people get to enjoy the spectacle of cherries, apples and other fruit trees in blossom which, in Japan, is revered as *hanami*.

Streuobstwiese is a German word that means a meadow with scattered fruit trees which, traditionally, provided fruit for rural communities. In Germany, these meadow orchards are now receiving state subsidies to be preserved because of their value as landscapes and habitats. Even if you don't live near one, you can follow the tourism guides to walk in them.

Orchard maths

- Some say an orchard can have 3 or more trees. Others say it has to be 5.
- Most fruit trees require 6–8 hours of sunlight a day in summer.
- Semi-dwarfing apple trees (up to 3.5m/10ft high) should be planted 3–3.5m (10–11ft) apart.

Malus domestica 'Cornish Gilliflower'

Orchard planting patterns

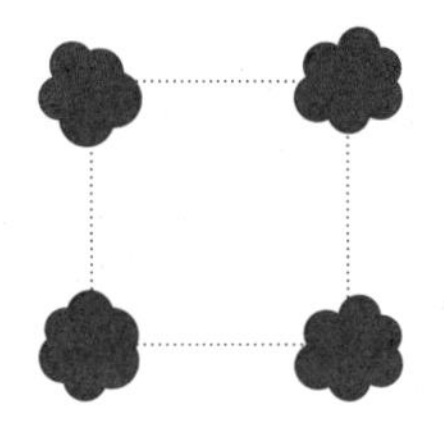

Square: Vertical-row planting pattern, usually north to south.

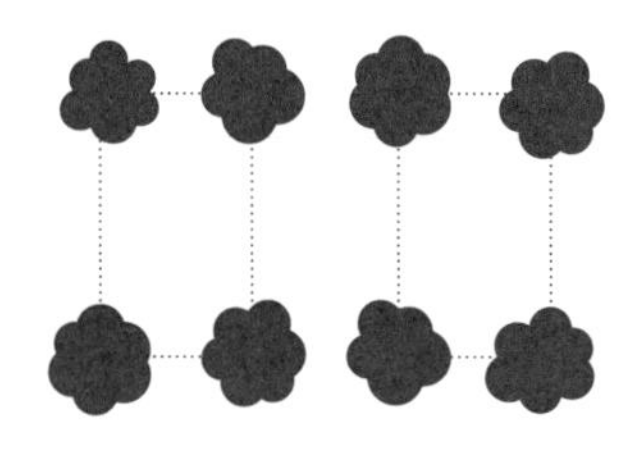

Rectangular: Vertical-row planting pattern.

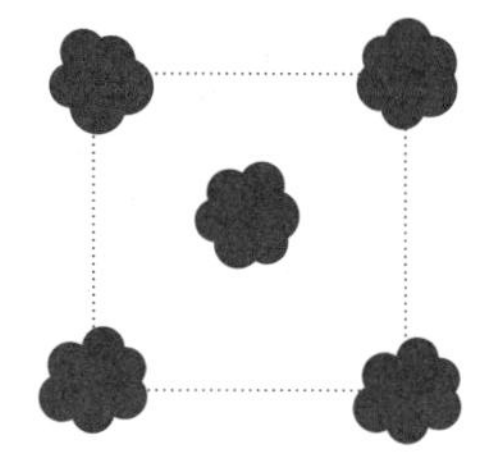

Quincunx: Trees planted in each corner of a square and one in the centre.

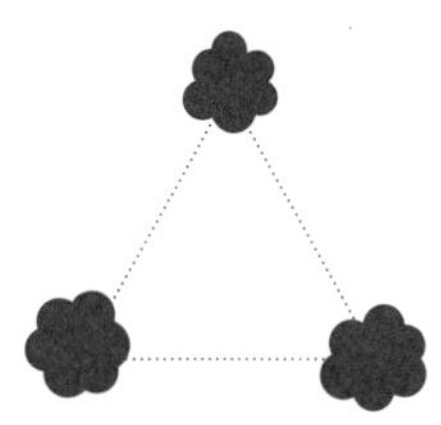

Triangle: One tree in each point of a triangle. If planted as a field, the trees are placed in vertical rows but with those in even-number rows planted midway between those in the odd-numbered rows.

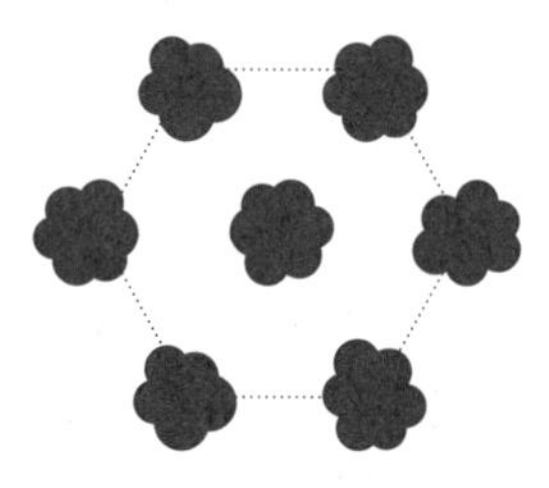

Hexagonal: Trees planted in each corner of a hexagon, with a seventh tree in the centre.

Contour system: Used for hilly terraced areas where trees are planted along the contour lines of the slope.

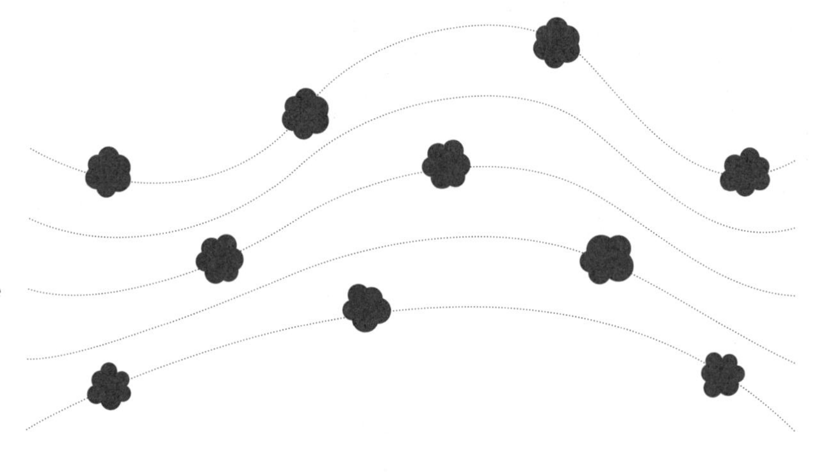

Plant a patio orchard

You don't need to have a field to have an orchard. A sheltered and sunny position on a patio or similar surface will do. Containers need to be stable and at least 45–50cm (18–20in) in diameter. Use good quality loam-based, peat-free organic compost and mix in some grit to aid drainage. You will need to water and feed the trees weekly in the growing season.

The trees: Most importantly, choose fruit that you love to eat. Then find their cultivars that are best suited to containers: they will be grafted onto semi-dwarfing or dwarfing rootstocks that limit the size and vigour of the tree. The shape is also important and choices for some include 'columnar', which are super-skinny, or fan and espaliers, which can be trained against a wall or with a free-standing frame.

Pollination: Some fruit trees are self-fertile, others will need a pollination pair to help things along. Fruit trees will be labelled with their pollination group (numbered to reflect when they flower). It's a good idea to place smaller containers, full of nectar-rich flowers and herbs to attract bees and other pollinators, near your trees.

'There are only ten minutes in the life of a pear when it is perfect to eat.'

American writer and philosopher
Ralph Waldo Emerson (1803–82)

Who has the biggest area of apple orchards?

1. China: 2,092,326
2. India: 313,000
3. Russia: 225,256
4. Turkey: 168,811
5. Poland: 168,811
6. Iran: 131,900
7. USA: 117,441
8. Uzbekistan: 111,575
9. Ukraine: 84,400
10. Pakistan: 75,230

Figures in hectares from Food and Agriculture Organization of the UN (2021)

Who produces the most apples?

1. China: 45.98m
2. Turkey: 4.49m
3. USA: 4.50m
4. Poland: 4.07m
5. Iran: 2.77m
6. India: 2.28m
7. Russia: 2.22m
8. Italy: 2.21m
9. France: 1.63m
10. Chile: 1.56m

2021 FAOSTAT, in metric tonnes

Maximize a small garden

Think big

- Get rid of clutter, which means everything – including pots with dead plants.
- Blur the boundaries and make use of views from your garden (the 'borrowed landscape').
- Pick your design style – formal or informal, tropical or herbaceous, sleek or rustic (see pages 46–50).
- Pick your colour scheme (see page 72).
- If you are going to use containers, the bigger the better.
- Limit landscaping materials to three types at most.
- Blend the house with the garden by matching landscape materials.
- Use vertical space with trellises and tepees.
- Create zones or 'rooms', if there is space for more than one.
- Create a sense of mystery with a path and a hidden destination.

Tricks of the eye

The 'slow reveal'

It's all about creating a journey that, however short or long the path, holds some mystery. This could be simply placing large-leaved plants in a way that obscures where a path leads. The slow reveal aims to create the opposite of a focal point, beloved in large gardens, which instantly reveals itself to be a bench or tree or sculpture. A slow reveal in a small garden creates the illusion that the garden is large enough to have unseen areas. What you don't want is a path that just ends: instead, create a seating area and place a water feature, bench or even just a chair.

Widen a thin garden

If your garden is narrow, then think about designing it on the diagonal. So, as opposed to a square, fill the space with diamond shapes or, even, one large diamond shape. It tricks the eye into thinking the space is wider than it is.

Acer palmatum and *A. negundo* 'Variegatum'

Lagerstroemia indica

Never too small for a tree

Small-garden trees include:

Acer griseum: Cinnamon-coloured peeling bark, white spring flowers, autumn colour.

Amelanchier laevis: White flowers in spring, autumn leaf colour, winter fruit.

***Sorbus* 'Joseph Rock'**: White flowers in spring, autumn colour, winter berries.

Lagerstroemia indica: Winter bark interest, bright pink or white flowers in summer followed by rich leaf colour in autumn.

***Acer palmatum* 'Sango-kaku'**: Japanese maple with bright coral-red bark and pinkish-yellow foliage turning green in spring then yellow in autumn.

Sorbus 'Joseph Rock'

Dahlia variabilis
Iris tectorum
and
Iris loptec
Rosa gallica
Syncarpha
eximia
Lilium
bulbiferum
Gentiana
acaulis
Coreopsis
verticillata
Aquilegia
vulgaris
Pyrus
communis
Convolvulus
tricolor
Smilacina
stellata
Lychnis
grandiflora

A question of colour

Light or dark?

Painting your garden boundary walls and/or fences can make a huge difference. Light colours, such as pale pink or white-green, will make the space seem bigger and brighter, but, if this is a sunny spot, it can also look washed out to the point of dazzling. Darker colours, such as deepest green or blue, will create an atmosphere that can work well, especially with lighting, and plants against a dark wall will pop.

Warm or cool?

What's the mood? If you want to feel energized, choose plants with flowers in hot colours such as reds and bright yellows. These will seem closer than they are. If you want to feel calm and relaxed, then go for cool tones such as blues, light purples and greys. These will seem further away. One way of making a garden seem bigger is to plant cool tones at the back and hot colours at the fore.

Wow! Or maybe not?

Do you want the wow factor? Research shows that you will need medium-size to tall plants with long-lasting flowers in a variety of colours. But perhaps, in a small garden, you might want to limit the colour palette to two or three hues, as a wide variety of colours can look cluttered and busy.

Is green a colour too?

A no-colour garden (i.e. green) allows you to concentrate on leaf texture and shape to provide variety. Colour psychology says our brains associate green with nature and healing.

Harmony or contrast?

This is all about mood. If you like the calm feeling that comes with colours that match and blend with one another, then you are after harmony. If you pick a dominant colour, such as yellow, then its harmonious colours will be on either side of it on the colour wheel. If, however, you want a bit of zing, then you are after contrast. The colours on the opposite side of the wheel are complementary without clashing.

Can colour create depth?

The idea of creating 'eye zones' in your planting can be done with colour (cool flowers to the back; hot to the front) and leaves (large leaves at the front; smaller, refined leaves to the back). Lighting and features such as water can also add depth. Place a few tall, airy plants near the front of a border. This allows you still to 'look through' the plant and creates a feeling of depth, which makes your garden seem larger.

The noble tomato

From deadly to delicious

For centuries, the tomato was feared in Europe. It was the Spanish who introduced the plant to Europe from the Americas in the early 16th century. But in northern Europe, the plant was viewed with suspicion, as it is related, and bears some resemblance to, deadly nightshade (*Atropa belladonna*, which is actually deadly). It wasn't until the mid-18th century that the 'poison apple' was accepted as not only edible but also delicious, and grown for more than ornamental purposes.

World-beating tomato

The tomato (*Solanum lycopersicum*) is the most popular vegetable in the world according to the UN Food and Agricultural Organization even though it is actually a fruit. Nutritionists see it as veg, botanists as fruit (which it is, anatomically). Others simply refer to it as a 'culinary vegetable'. Native to South and Central America, its three largest producers are China, India and the USA. So, which veg won second-best? Don't cry, but it's the onion.

Anatomy of a tomato

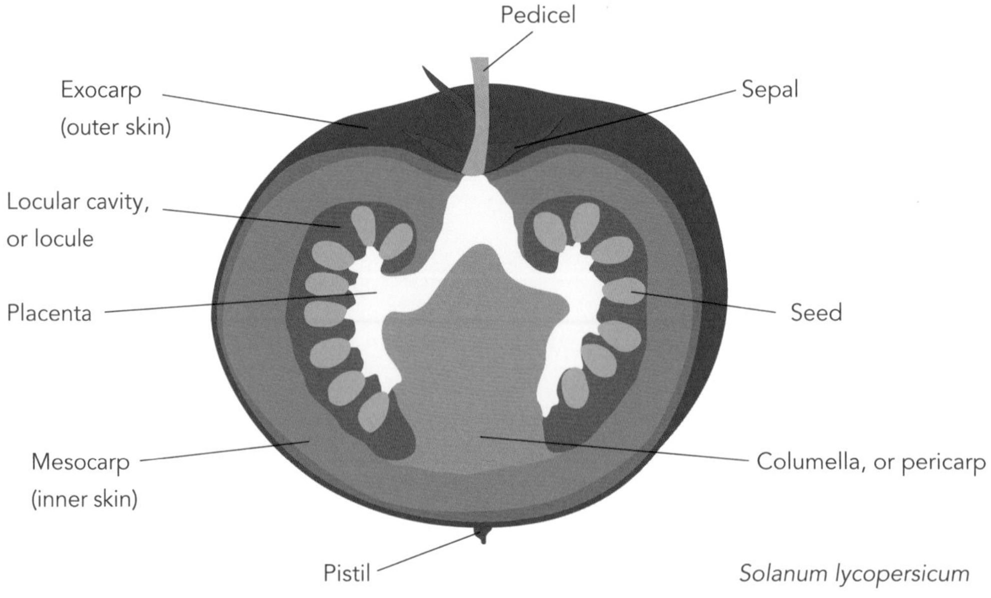

Solanum lycopersicum

The humble veg plot

It takes all sorts of ways

Allotments: A collection of plots gardened separately by individual plotholders.

Raised beds: Enhanced drainage, good for weed control; ideally not more than 1.2m (4ft) wide if against a boundary.

Plot: Planted in straight rows running north to south.

Terraced garden: Level sections created on a slope, following the contours.

Formal squares: Garden divided into geometric patterns, including four-square, with each square divided again into four.

Potager: A kitchen garden that is both attractive and productive with flowers, vegetables and herbs grown together.

Guerrilla: Surreptitious, technically illegal, cultivation of public and private land.

Community garden: One garden worked by a group of local people.

Permaculture: Sustainable and self-sufficient method that mimics natural ecosystem.

When to plant

Following nature: phenology
Forget those instructions on seed packets. Here's another way to know when to sow your vegetable seeds: phenology. This is the art of observing nature for signs that now is the exact right time to plant or, as it is sometimes called, old wives' tales. *The Old Farmer's Almanac*, published in the USA since 1792, has always been keen on linking the likes of migrating birds or other natural signs to sowing time. So remember to plant corn when the oak leaves are the size of a mouse's ear; plant peas when forsythia blooms; and wait for the lilac flowers to fade before sowing cucumbers and squash.

Following the moon
The idea behind moon-based planting is that the moon is in charge of all things watery on Earth, including sap in plants. There are many variations on this, but the basic plan is to sow or plant between the new moon and first quarter (the light of the moon phase). When the moon is waning, or diminishing (the dark of the moon phase), is the best time for pruning and weeding. If there is a drought, plant just before the full moon, apparently seen as a particularly wet time.

Make a twiggy cage

A twiggy cage is an attractive alternative to netting to keep out the birds. Use twiggy branches of hazel or birch if possible (though pruned twiggy branches from large shrubs will also work) and push them into the ground. Pull the flexible smaller twigs on top towards one another, weaving them together to form a natural canopy.

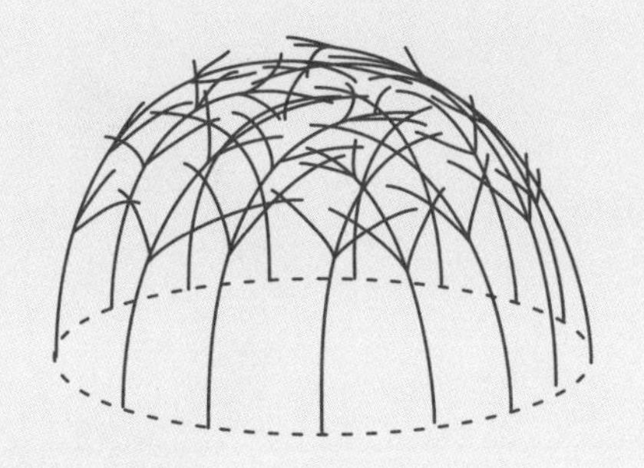

Crop rotation

The idea of rotating crops is all about refreshing the soil and dissuading diseases and certain pests by growing different families of vegetables. In a smaller plot, mixed plantings will serve much the same function. A classic four-part crop rotation would be to plant:

Year 1: Legumes, which fix nitrogen.

Year 2: Brassicas, which need nitrogen.

Year 3: Potatoes, which break up the soil and keep the weeds at bay.

Year 4: Root veg and onions. Then start all over again.

Manihot esculenta

Flavours from South America

- Lemon drop pepper *Capsicum baccatum* 'Aji Limon': Peruvian hot pepper
- Cassava, or yuca (*Manihot esculenta*): Mild, slightly nutty root vegetable
- Ulluco (*Ullucus tuberosus*): Root veg akin to a mix of beetroot and potato
- Papa púrpura: Purple potato from Peru
- Tamarillo (*Solanum betaceum*): Egg-shaped tomato-like fruit

Veg families

A family affair

Gardening is a family affair and never more so than in the vegetable patch. All plants are classified into families and, in general, they share physical characteristics. These are the main families that you'll find in your veg garden. As always, it's good to know who is related to who. Often, crop rotation is done by family. Plus, knowing the families helps you notice similarities – and occasionally find the rogue element too.

FAMILY: BRASSICACEAE

Brassicaceae is known as the mustard, crucifer and cabbage family. These plants originated in the Eocene (from about 56 to 34 million years ago) in the Irano-Turanian region and spread worldwide. Sanskrit records show they were grown in India as far back as 3000 BBCE. They are now a highly diversified and economically important family that includes oilseed (rapeseed and mustard) and vegetables such as kale and broccoli, hailed for their health-giving properties. Other, perhaps less popular, family members include turnips and Brussels sprouts.

FAMILY: AMARYLLIDACEAE

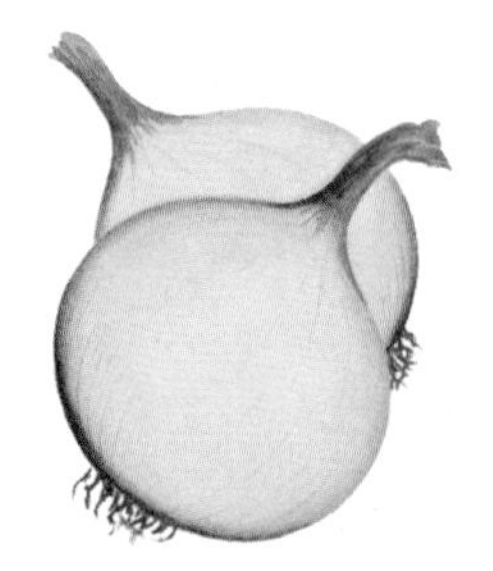

The family Amaryllidaceae is known for being perennial and mostly bulbous. They are monocots (most vegetables are dicots) and found in all areas of the world. Most possess strap-shaped leaves. Flowers include daffodils, snowdrops and amaryllis. Vegetables in the family include garlic, leeks, chives and onions.

FAMILY: CUCURBITACEAE

Cucurbitaceae includes some 800 species that are known as the gourds or cucurbits. In general, they grow best in warmer parts of the world and are economically important. Favourites include cucumbers, melons, watermelons, pumpkins and squash.

FAMILY: SOLANACEAE

The Solanaceae is known as the nightshade family. Many of them contain solanine, an alkaloid that functions as an insecticide while the plant is growing. The family includes the likes of deadly nightshade (*Atropa belladonna*, see page 74) and tobacco. Vegetables include warmth-loving tomatoes, peppers, chillies and aubergines. Potatoes are also in this family, though they don't care as much about warmth. Tip: When potatoes go green, it's because the solanine has increased, and it means they are poisonous.

FAMILY: FABACEAE

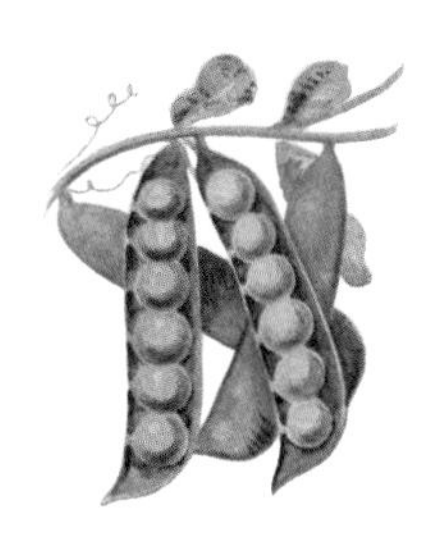

Plants in the Fabaceae family are an essential part of the veg garden. All members have 'fruits' (as they are called) or legumes, which include the allotment stalwarts beans and peas. All have seeds held in a pod. Others include chickpeas, lentils and soya beans. Legumes often form a symbiotic relationship with nitrogen-fixing soil bacteria and so-called green-cover crops, which include red clover (*Trifolium pratense*) and alfalfa (*Medicago sativa*).

FAMILY: ASTERACEAE

What a popular family this is. It contains more than 20,000 species of herbs, vegetables, shrubs, flowers, vines and trees – all flowering. Plants in the family are found all over the world and include the beloved daisy (*Bellis perennis*) as well as aster, chrysanthemum, thistle and dahlia. Veg includes lettuce, chicory, sage, tarragon, artichokes and sunflowers. The flowers of Asteraceae are in the classic daisy shape.

FAMILY: APIACEAE

Apiaceae comes from the word *apium* which was originally used by Pliny the Elder (in about 50 CE) for a celery-like plant. The plants tend to have hollow stems and an aromatic scent, with flowers arranged in an umbellifer (umbrella-like, see page 12) inflorescence. The family includes celery, coriander, parsley, fennel and carrots. Sea celery (*Apium prostratum*) is native to Australia and New Zealand, and was embraced by Captain Cook and his crew on the *Endeavour* to help prevent scurvy.

The dirt on soil

You'll need a worm

If you want fertile soil, then it helps to have a worm or two, or maybe a couple of thousand. It sounds a lot, but studies show that even poor soil can harbour thousands per acre, and rich soil a million or more. They may be armless and legless hermaphrodites without a skeleton, but worms are considered to be ecosystem engineers that improve soil quality by adding nutrients and providing aeration. They have no eyes but can sense vibrations, which Charles Darwin discovered by placing them on a piano and, as he played, watching them start to burrow.

Slippery, soapy or gritty?

Clay: Dense and heavy, made up of tiny particles that hold onto nutrients and moisture. Slow to warm in spring. Drains poorly; cold and wet in winter, dry and hard in summer. Smeary to the touch and sticky when wet. Add organic matter to alleviate these problems.

Sandy: Light and airy, made up of large particles, which means it drains too freely and doesn't hold on to moisture, nutrients or organic matter. Often acidic. Feels gritty to the touch. Add organic fertilisers and organic matter if needed, but be aware that sandy soils lose more nutrients and carbon over time compared to other soils.

Silty: Made up of medium-sized particles that are fertile and relatively light. Drains faster than clay. Easily compacted. Feels somewhat soapy and slippery. Add organic matter to alleviate these issues.

Loam: A mix of clay, sand and silt. Good to work with. Will probably be either mostly clay or sand, though.

Peat: Mainly organic matter. Very fertile and moisture retentive. Slightly acidic. Seldom found in gardens.

Chalky: Lime-rich soils, mostly made up of calcium carbonate. Alkaline.

Right plant, right place, right soil

Plants have varying preferences when it comes to soil. Soil pH – whether your soil is acid, alkaline or neutral – has a lot to do with this as it will affect the availability of nutrients in the soil. A test kit will give you results within minutes. The science behind this rests on an equation: $pH = -\log[H+]$

The 'p' comes from *potenz*, the German word for 'power', or 'potency', and the 'H' is for hydrogen; so the equation is an abbreviation for the 'power of hydrogen'. It's an inverse indicator, so the more hydrogen ions in the soil, the more acidic it will be and the lower its score. Pure water is pH7, which is neutral.

Acid lovers	Camellia
	Blueberry
	Azalea
	Heather
	Magnolia

Alkaline lovers	Clematis
	Dianthus
	Campanula
	Geranium
	Lavender

The 'father of soil science'

Russian geologist Vasily Dokuchaev (1846–1903) was the first person to see soil as a living system and as a biological science. He saw that it was not dead rocks but the basis of a living system formed as a result of climate, organisms, topography, time and, of course, the parent rock. He even has a crater on Mars named after him and also the Ukrainian city of Dokuchaievsk.

Pelargonium

Take up your tools

Classic gardener toolkit

1. Hand fork, for planting and weeding
2. Hand trowel, for planting and weeding
3. Dibber, for making holes
4. Planting string line, for keeping planting rows straight
5. Twine and labels, for tying in and labelling plants
6. Soil riddle or sieve, for creating fine tilth for planting seeds
7. Secateurs, for deadheading and pruning
8. Watering can
9. Comb rake, for levelling and clearing soil, making furrows for seeds
10. Lawn rake, for aerating lawn and collecting leaves
11. Draw hoe, for hoeing between plants
12. Dutch hoe, for weeding and clearing
13. Spade, for digging, planting and breaking up soil
14. Fork, for digging and breaking up soil
15. Shears, for trimming and pruning
16. Wheelbarrow
17. Lawnmower (if you have a lawn)

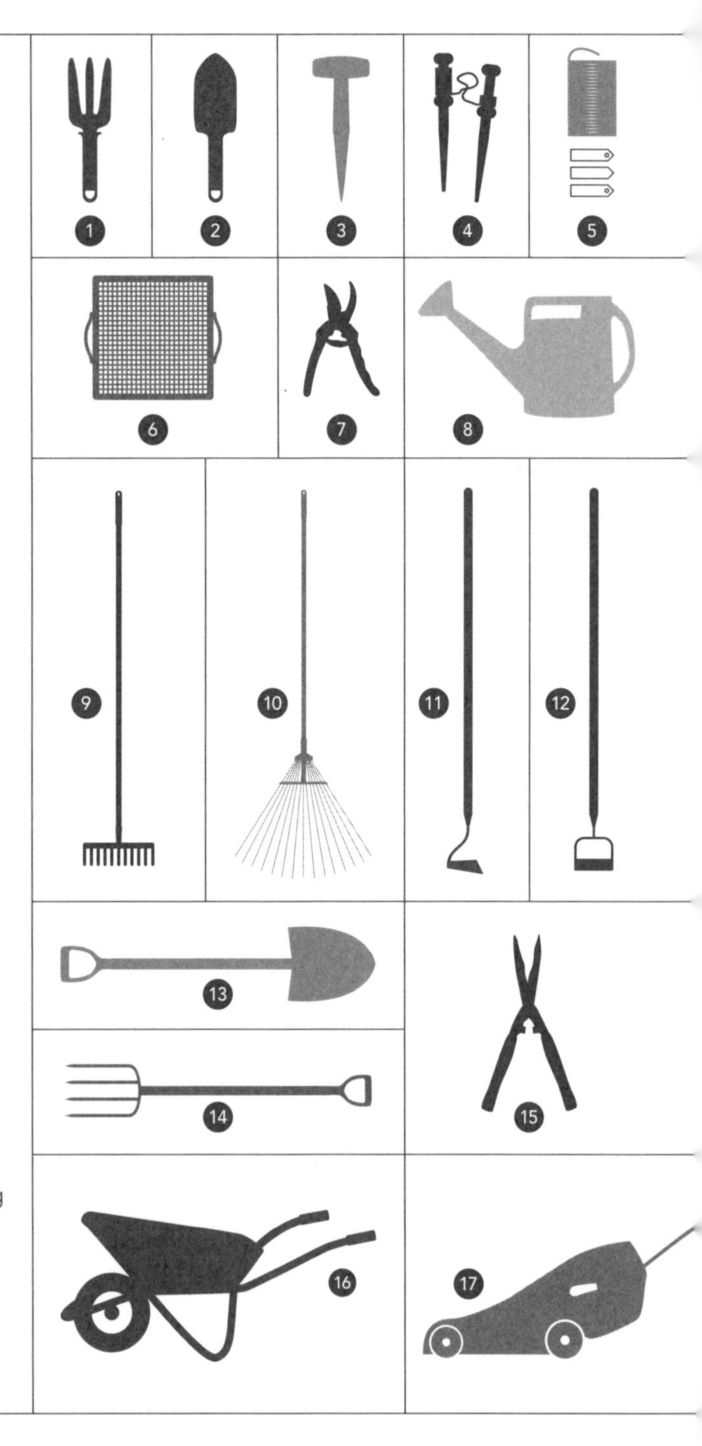

The ingenious thumb pot

The thumb pot was used in Europe from the 13th to the 17th century for the tricky task of gently watering of seeds and young seedlings. These earthenware pots, about the size of a small watering can today, had small holes in the base and a tapered neck with an end as large as your thumb. The pot would be immersed in water, filling from the bottom. The flow of water would be controlled via the thumb. Surely it's time to bring these back.

'He that would perfect his work must first sharpen his tools.'

Chinese philosopher Confucius (551–479 BCE)

Strange but true tools

Cucumber straightener: As it says on the tin, made of glass, 1880s.
Fern trowel: Long and thin for digging between rocks.
Dandelion weeder: Long and thin to get those roots.
Spiked snowshoes: To be worn by humans for aerating lawns.
Tulip trowel: Dutch, of course, with a jagged end.

The 'Song of the Hoe'

The hoe that we keep in our garden sheds may seem entirely ordinary to us, but to the Sumerians, who lived from 4000 to 1700 BCE in Mesopotamia, in what is now southern Iraq, it was an extraordinary tool that transformed their lives. They saw it as a godsend, literally, and the 'Song of the Hoe' is a Sumerian creation myth that describes how the ancient god Enlil uses his hoe (made of gold with a lapis lazuli blade) to create daylight and to make a brick mould to create civilized man. Enlil goes on to make many hoes so everyone can get to work cultivating the world.

Tools from Japan

Hori hori: This translates as 'to dig' in Japanese. A doubled-edged knife with one side serrated and the other a smooth, scalloped blade. Used for weeding, transplanting, cutting, digging and scooping.
Bonsai scissors: Small scissors for shaping bonsai.
Shuro brush: Small brush made from palm fibres. Used for sweeping around bonsai and other potted plants.

Magical gardens

By the light of the moon

How to create a night garden? We all aspire to greatness and so may think of the Moonlight Garden at the Taj Mahal in India, which reflects its beauty in a large pool. Take this as inspiration to play with reflection, sound and scent; muted lighting; soft soundscape; scented plants that draw night-time moths and smell delicious – try honeysuckle, woodland tobacco plant (*Nicotiana sylvestris*), stock and jasmine. Use pale flowers such as a climbing white clematis. Don't forget a softly lit seating area – perfect for star-gazing.

'If you look the right way, you can see the whole world in a garden.'

British-American author Frances Hodgson-Burnett, *The Secret Garden* (1911)

The Taj Mahal, India

Alice in Wonderland style

- Flamingo statues that double as croquet bats
- Croquet balls styled as hedgehogs
- People dressed up as playing cards
- White roses and red paint
- A queen
- A very large grin that, when the mist clears, belongs to a cat

Edward Scissorhands style

If you want to emulate the whimsical garden from the 1990 film, then you'll need topiary everywhere, placed near one another and shaped like people, animals and objects of all kinds. Handy if you have cutting-edge hands, but most of us will need the garden shears.

Alice and her 'croquet bat'

Can you keep a secret?

The idea of a secret garden undoubtedly has a magic to it, with shades of the forbidden, the magical and the ultra-private. To create one requires imagination and thinking outside the box (hedge and cardboard varieties). It's key that your secret place is different from the rest of the garden. If you're ambitious, dig down to make a sunken area. It helps to have a secret entrance, of course, which could be a *trompe l'oeil* gate (cover it with ivy). Think archway, double hedge, or just a little narrow squeeze-gate. There, you're in! Fill your space with plants and items that delight: comfy chairs, pebbles, water, rustling grasses, scented blooms, fantastical shapes. Don't tell anyone about it, though: it's a secret.

Gnomes and fairies

Germany's 'little folk'

The humble garden gnome, both beloved and disliked by probably an equal number of gardeners around the world, has quite a history. Typically, they are seen now as humorous figures, male dwarfs in red pointy hats, who add a bit of fun to the garden. But they hark back, with their rosy cheeks and white beards, to the Black Forest in Germany in the 1800s. They were seen as 'little folk', really a kind of fairy, who helped out in the mines and on the farm. Both world wars saw production and popularity wane but, in the 1970s, there was a resurgence of garden gnomery. Now, apparently, there are 25 million gnomes in Germany alone.

No place like gnome

Even in a country such as England, which is known for its eccentrics, Sir Charles Isham (1819–1903) stands out as particularly unusual. Isham was a rural improver and spiritualist, who spent 50 years from 1847 cultivating his garden at Lamport Hall in Northamptonshire. It is his rockery that made him famous, though: a miniature mountain world populated by small figures who represented the *Berggeister*, or mountain spirits, of Germany. They are an amazing lot: some are miners who are on strike, others hold placards with demands, while others loll around in a leisurely mode. Isham did not see his gnomes as imaginary or mythological creatures: he believed they had real lives and were occult phenomena. Thus, the gnome had come to Britain.

'Isn't it enough to see that a garden is beautiful without having to believe that there are fairies at the bottom of it too?'

Douglas Adams, English author of *The Hitchhiker's Guide to the Galaxy* (1978)

The Fairy Queen

Fairy garden must-haves

It's all about trying to entice a good fairy into your garden, to give you good luck, and it carries on a folkloric tradition found throughout Germany, France, Scandinavia, Britain and Ireland. Here's what you'll need:

- Moss to dance on
- Des-res home in tree
- Smart front door
- Winding pathways
- Tiny stepping stones
- Mushrooms
- Ponds and, preferably, a water wheel

Xerocomus communis

Seeds of survival

Seeds to celebrate

What plants did enslaved Africans introduce to the Americas after being forced to leave their homelands? What did they grow in their domestic gardens? How were the plants used in cooking and traditional medicines? These are some of the themes explored in the exhibition Seeds of Survival and Celebration at the Cornell Botanic Gardens in upstate New York. The multitasking plants in question have been used for sustenance and healing, and have become part of modern life too.

Black-eyed pea (*Vigna unguiculata*)
This is not a pea but a bean, which is native to West Africa. These legumes provided food for millions of Africans on slave ships and later, in the USA, were planted in their own vegetable gardens. A traditional African American dish called Hoppin' John uses black-eyed peas with smoked ham hocks or bacon, onions, spices and collard greens (spring greens): this was seen as bringing good luck and prosperity on New Year's Day.

Okra (*Abelmoschus esculentus*)
A vegetable used in African cooking for thousands of years, okra was taken across the Atlantic on slave ships and grown in domestic gardens. It's the ultimate in versatility and can be stewed with tomatoes, battered, fried, pickled or simmered in stews. Enslaved people ate okra stewed in chicken broth to counter chills, and the flower blossoms were applied to boils.

Capsicum annuum

Collard greens (*Brassica oleracea*)
This leafy green vegetable was grown by enslaved people in their own gardens. The leaves were simmered with the likes of bacon or ham hocks. They were also used to treat boils and pleurisy.

Sorghum (*Sorghum bicolor*)
This cereal crop was first domesticated in Sudan thousands of years ago. Sorghum was grown in America for use in cakes, bread and porridge. It was also used to sweeten medicines.

Pearl millet (*Pennisetum glauca*)
A staple food crop in Africa, pearl millet was planted around the living quarters of enslaved people. The millet flour was used for cakes and bread.

Hot pepper (*Capsicum frutescens*)
A pepper that has grown wild in Africa for centuries and is now cultivated commercially there. A key ingredient in the hot sauce peri-peri, which comes from the Swahili *piri-piri*.

A very unusual tree

A bottle tree is a centuries-old tradition that comes from parts of Africa, including the Democratic Republic of the Congo. Bottles are hung or balanced on bare branches of trees as a way of redirecting evil spirits. Found objects, including chairs and pipes, often decorated gardens as a way of attracting ancestral spirits. Bottle trees were introduced to the Americas by enslaved Africans. They often feature blue glass and can be seen in gardens throughout the American South.

'Cotton was a main cash crop but it was also used by pregnant enslaved women to help induce labour. Cotton is the plant that was most harmful to these people and yet, they turned it into something that was useful for them.'

Jakara Zellner, Garden Ambassador Co-lead, Cornell Botanic Gardens' Learning by Leading programme

Musa 'Dwarf Brazilian'

Chapter 3

Noteworthy Nature

The natural world makes gardening and landscape design look effortless. Here is an exploration of the, at times, mysterious way that plants, from small weeds to mighty trees, interact with one another and the wider world, creating intricate habitats and fascinating landscapes.

The birds and the bees

What are pollinators?

A pollinator carries pollen from flower to flower, or within a flower, to enable fertilisation. So, who are the pollinators?

- Bees and wasps
- Beetles
- Butterflies
- Hummingbirds and other birds
- Moths
- Hoverflies and other flies
- Mice
- Lizards, geckos, skinks
- Tiny crustaceans such as woodlice
- Tropical bats and other mammals
- Humans (when armed with a very small paintbrush) and drones
- Wind and water
- And, of course, the self-pollinators

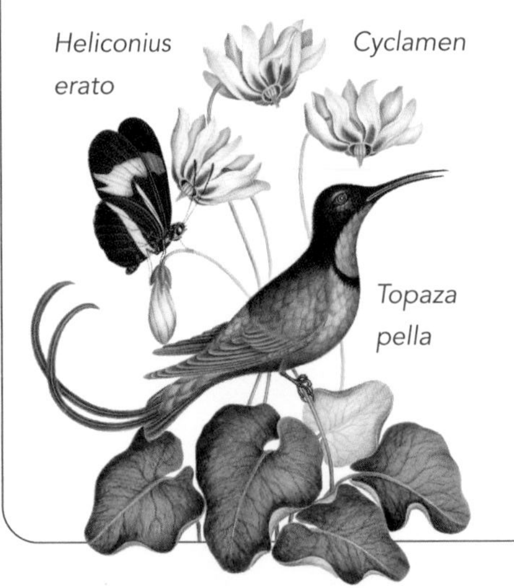

How bees see the world

To see the world through the eyes of a bee is to remove the idea of 'red'. And that's just for starters. While we see colours on a spectrum of red, green and blue, they see blue, green and ultraviolent light instead. There is even a colour called 'bee's purple', which is a combination of yellow and UV light (which we, of course, can't see). This makes their view of a flower border an almost psychedelic experience, with flowers that glow. Many flowers reflect a UV light that appears to a bee like a landing strip, guiding it to the pollen and nectar. Good for the bee, good for the flower. Now that's symbiosis.

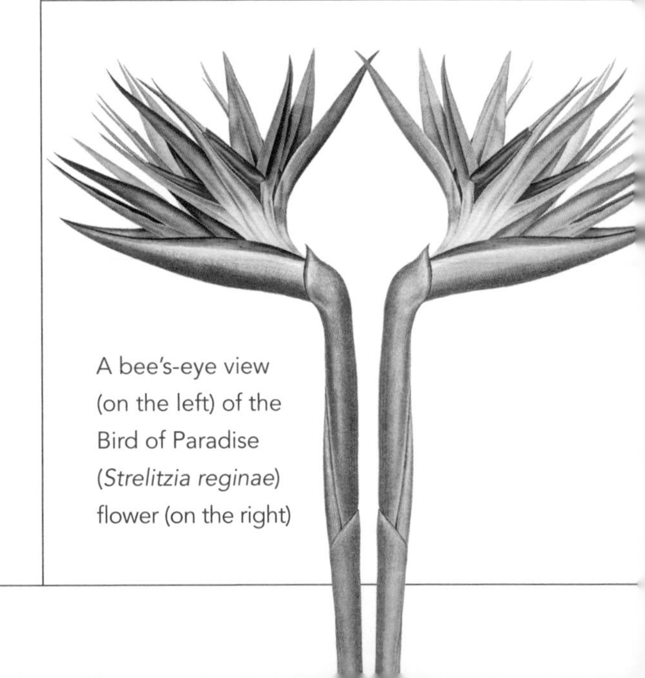

A bee's-eye view (on the left) of the Bird of Paradise (*Strelitzia reginae*) flower (on the right)

One of the world's largest bumble bees, *Bombus dahlbomii*, shown here actual size

The buzz

Bumblebees have a sort of dance when it comes to pollination. A female bumblebee can vibrate its body to dislodge the pollen from a flower, then comb the pollen off its hairy body into little saddlebags on its legs called corbiculae. This is called, appropriately, buzz pollination. Most of the pollen is destined for the nest to feed the bumblebee larvae but some will transfer to the next flower.

Save the pollinators!

- Have plants in bloom for every season.
- Plant night-scented flowers for moths.
- Plant in groups of three or five in drifts.
- Use many different types of flower.
- Herbs and wildflowers are pollinator favourites.
- Choose varieties with a single ring of petals as it's easier for bees to access the nectar and pollen.
- Put out a saucer of water with pebbles.

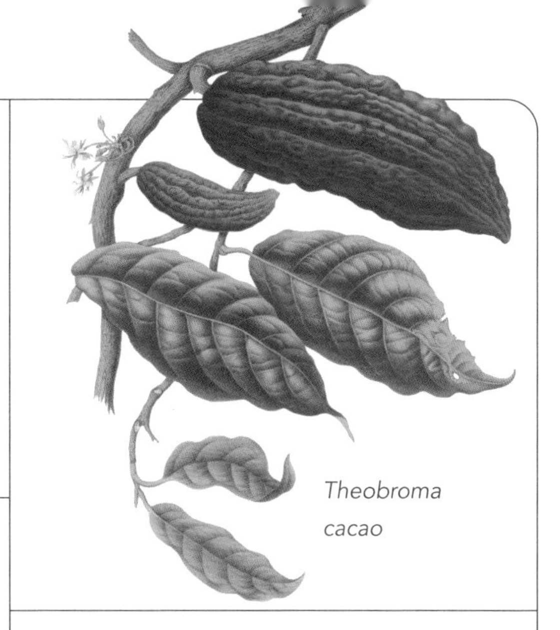

Theobroma cacao

Cacao: a tricky customer

The cacao tree (*Theobroma cacao*) seems to have gone out of its way to make things difficult for itself. Its little pink and white flowers emerge directly from the trunk. Each petal curves into a hood that fits down around the male pollen-making structure. It's all so tiny that there is no point getting something as large as a bee involved. Instead, tiny flies known as chocolate midges (*Forcipomyia* spp.) crawl into the hood, though once they get there, scientists aren't quite sure what they do (there's no nectar). They emerge and then must find another cacao tree to pollinate. No wonder that in many places pollination rates are pitiful: it is estimated that just one in 400–500 flowers (which only last for a day or two in any case) will produce fruit and only 10–30 per cent of pods will reach maturity. Makes you appreciate that chocolate bar all the more, doesn't it?

Plants in harmony

10 companions for roses

1. Lavender (*Lavandula*): Bee magnet with attractive leaves that will hide lower rose stems.

2. Toothpick plant (*Ammi visnaga*): Feathery foliage and green and white domed flowers that contrast beautifully.

3. Chives (*Allium schoenoprasum*): Onion family members are said to ward off aphids and reduce fungal diseases.

4. Mexican feather grass (*Stipa tenuissima*): Delicate grass that will add movement and structure.

5. Balkan clary 'Caradonna' (*Salvia nemorosa* 'Caradonna'): Dark purple spikes provide striking background contrast.

6. Daisies such as *Argyranthemum* 'Grandaisy Ivory White': Cheerful underplanting that will also hide basal stems.

7. Purple top (*Verbena bonariensis*): Tall and elegant with purple flowers that add grace and interest.

8. Common yarrow (*Achillea millefolium*): White umbels and feathery foliage that may attract ladybirds that helpfully consume aphids.

9. Feathertop (*Pennisetum villosum*): Rabbit-tail flower spikes add movement.

10. Tulip (*Tulipa* spp.): Dense planting with contrast colour will add early-season interest.

Rosa indica

Tulipa

Best plants for pollinators

- **Aubretia**: Scented flower carpet provides early season nectar and pollen.
- **Grape hyacinth** (*Muscari*): Early bright blue spikes attract pollinators like the hairy-footed flower bee (*Anthophora plumipes*), one of the earliest to emerge from hibernation.
- **Tobacco plant** (*Nicotiana* spp.): Strong evening scent attracts night-flying moths.
- **Honeysuckle** (*Lonicera* spp.): Easy-to-grow climber adored by long-tongued pollinators.
- **Mallows** (*Malva* spp.): Musk and common mallows are particularly attractive to bumblebees and honeybees.
- **Sea holly** (*Eryngium* spp.): Blue-flowered eryngium is a magnet for most bees and butterflies.
- **Single sunflowers** (*Helianthus* spp.): Late-season favourite.
- **Butterfly bush** (*Buddleja* spp.): Beloved by adult butterflies. Grow a few varieties to extend the flowering season.
- **Ivy** (*Hedera* spp.): Important source of nectar and pollen in the autumn for bees, hoverflies and wasps.

Hedera helix

Lonicera

Rewilding on a large, and small, scale

Wetland managers

Beavers are rodents, which doesn't sound promising in terms of being able to transform a landscape. Yet, in ecological terms, they are ecosystem engineers and a keystone species. Their dams alter rivers and streams, creating extensive wetland habitats that reduce pollution and wildfires. One study in the Adirondack Mountains, USA, found that their presence increased plant diversity by a third. When beavers were reintroduced in Devon, UK, they tackled the dense willow canopy, reinvigorating the wet and rushy pasture of the Culm grassland. But beavers are not welcome everywhere: they are native to the northern hemisphere and banned in countries such as New Zealand.

Castor canadensis

A sting in your wild garden's tale

Here's one small step towards a wilder garden: allow a patch of stinging nettles to grow. This rather painful hero weed, *Urtica dioica*, is a wildlife pitstop. It's the food plant for the caterpillars of the red admiral (*Vanessa atalanta*), small tortoiseshell (*Aglais urticae*), painted lady (*Vanessa cardui*) and comma (*Polygonia c-album*) butterflies. Ladybirds will arrive to eat the aphids sheltering among them and seed-eating birds will be feasting in the autumn. Nettles also make an excellent natural fertiliser, and their stalks (once the stingers have been removed) can be twisted into a twine.

Vanessa
atalanta
Aglais
urticae
Urtica dioica

It's all about trophic cascades

Trophic cascades is the ecological term that is used to describe how rewilding affects the environment. It refers to the knock-on effect of changes that trickle through the trophic levels (the hierarchical categories of the food chain within an ecosystem). The most famous example is the introduction of wolves in America's Yellowstone National Park which began to keep elk populations under control, which meant less grazing and resulted in riverbanks regreened with trees and shrubs. Plants are producers and are on the first trophic level: they will be affected by, and will affect, rewilding in myriad ways. A trophic cascade can be as grand as the wolves of Yellowstone or as small as a new garden wildlife pond that attracts frogs that then eat slugs, which consequently eat fewer leaves of your plants.

Give your lawn a break

If your lawn is one of those that has been cut regularly, every Sunday for years, you are likely to have created the perfect habitat for several species of wildflower, including those that can live for years without flowering, their rosette of leaves forming close to the ground. They are also called 'lawn weeds' and include daisies (*Bellis perennis*), ribwort plantain (*Plantago lanceolata*) and cat's ear (*Hypochaeris radicata*). They will also include creepers such as selfheal (*Prunella vulgaris*) and speedwell (*Veronica repens*). If you take a mowing holiday, then sit back and watch your lawn turn into a colourful wildflower meadow.

You don't need a wildebeest to rewild, but it helps . . .

The blue wildebeest (*Connochaetes taurinus*) is the lifeblood of the Serengeti, a 30,000 square-kilometre (11,500 square-mile) grassland in Tanzania, for it is they that lead, from July to October, what is called the Great Migration. This is the movement of 2 million blue wildebeest and 700,000 other grazing species across the plain, looking for food. For centuries this kept the Serengeti healthy, functioning as a huge carbon sink (see page 108). In the mid-20th century, viruses from livestock reduced the numbers of wildebeest to only 300,000, causing the plain to fill with choking ground vegetation, which contributed to wildfires. Instead of absorbing and storing carbon the Serengeti became a producer of greenhouse gases. Enter disease management, which helped the wildebeest population to recover, restoring balance to the plain and, in less than a decade, allowed it to once again function as a vital carbon sink.

Rhododendron ponticum

The bison are back!

And that is good news for grasslands. The American bison (*Bison bison*) and its European cousin (*B. bonasus*) were hunted to near extinction. In Europe, they have now been reintroduced in Germany, Switzerland, Poland, Belarus, Lithuania and, in July 2022, in the UK as well. In America, the animal seen as a national symbol is home on the range once again. Bison are big browsers of bushes, bramble and trees, which keeps open land, well, open. In the UK, within hours of their release in an ancient wood near Canterbury, the bison were thrashing and stomping on the invasive *Rhododendron ponticum*.

Bison bison

Slugs and plants

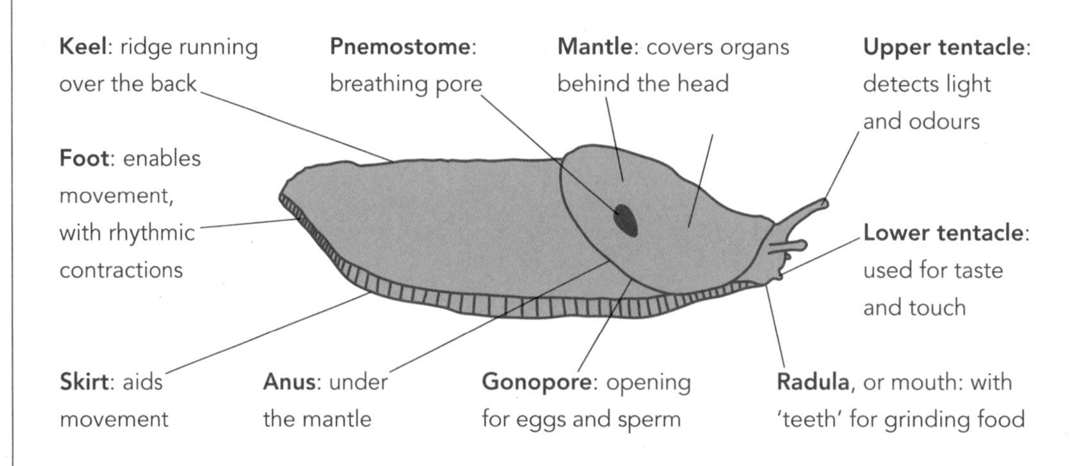

Sluggish facts

- Slugs are hermaphrodites (that is, they possess both male and female sexual organs) but often mate anyway.
- Each slug produces about 30-plus eggs several times a year.
- They have thousands of microscopic teeth that are constantly growing and being replaced.
- They 'smell' with their upper tentacle and use this to find food.
- They can live to be several years old.
- They produce a scented slime trail that helps them retrace their 'steps'.
- They are slow travellers, with the fastest (*Deroceras invadens*) clocking just 0.01mph (0.016km/h).
- They can stretch to 20 times thinner to squeeze into narrow gaps.

Pest or eco-warrior?

Slugs have had a very bad press for a very long time when it comes to gardeners. This is because they like to munch our plants, reducing the likes of hosta leaves to a series of holes held together by green threads. But can we really blame slugs for eating? If you look at it another way, they are an important part of the ecosystem. Yes, they eat some leaves that we'd love to stay intact, but mostly they feast on decomposing matter. In this way, they can be seen as eco-warriors: fanatical recyclers that excel in clearing out dead matter and, as is the way in the natural cycle of life, are themselves a food source for birds, frogs and hedgehogs.

Slugging it out

You can still think of most slugs that we find in our gardens as eco-warriors while wanting to protect your plants from them. Let's take a leaf, so to speak, out of the wisdom of ancient Chinese general and philosopher Sun Tzu, who said we should all know our enemy. Slugs are relatively easy to figure out. They tend to feed at night, spending the day in moist, dark places. They are voracious eaters, consuming up to 40 times their weight daily. We know that, come dusk, they will be out and about, looking for food.

Before the battle begins, it is important to remember that you will never get rid of slugs entirely. They are hermaphrodites that reproduce many times a year. No matter what you do, there will be slugs in your garden. They have the advantage of numbers.

So, how to outsmart such an enemy?

Search and remove

Go out at night with a torch and look in all the obvious places (under leaves etc). Research has shown that you must move a slug some 20m (66ft) away for it to not 'come home'. The often-deployed method of throwing slugs into your neighbour's garden is also not a friendly way to behave. Another idea would be to relocate slugs to the compost heap, where they can do good, munching away.

Green shield

Slugs are believed to dislike some plants, such as mint, chives, garlic, geraniums, foxglove and fennel. Plant these around particularly vulnerable plants (such as lettuce).

Horticultural 'rings of steel'

- Soot mixed with wood ash and lime
- Eggshells and seashells with sharp edges
- Sand, gravel, grit
- Crushed nut shells
- Pine needles
- Hair, human or pet; sheep's wool
- Sawdust

Remember that these things must have worked at one time or another as they have entered slug folklore, but some studies (and many gardeners) may also tell you that they aren't effective barriers.

Underground resistance

Add the nematode *Phasmarhabditis californica* to the soil, which can control slugs for some weeks.

The incredible life of a tree

The oldest trees in the world

Non-clonal trees

Methuselah, Great Basin bristlecone pine (*Pinus longaeva*), White Mountains, California, USA: 4,855 years old.

Alerce Milenario (aka Gran Abuelo or Great Grandfather), Patagonian cyprus (*Fitzroya cupressoides*), Alerce Costero National Park, Chile: at least 5,000 years old.

Jaya Sri Maha Bodhi, Sacred fig (*Ficus religiosa*), Anuradhapura, Sri Lanka: claimed to be from the bodhi tree under which Buddha was enlightened. Oldest living human-planted tree: 2,311 years old.

Clonal trees

Antarctic beech (*Nothofagus moorei*), Springbrook National Park, Queensland, Australia: three trees about 2,000 years old are parent trees to others in the forest. Clonal age: less than 70 million years.

Wollemi pine (*Wollemia nobilis*), Wollemi National Park, New South Wales, Australia: thought to be extinct until found in 1994. Clonal colony: 60 million years old.

Pando, Quaking aspen (*Populus tremuloides*), Fishlake National Forest, Utah, USA: 43-hectare (107-acre) forest with about 47,000 stems, continually renewing, with individual stems up to 130 years old. The heaviest known organism in the world, weighing 6,000 tonnes. Clonal age: 14,000–80,000 years old.

Non-clonal or clonal?

Clonal trees are genetically identical and share a root system. There can be thousands of trees in a clonal colony and they are considered a single organism. Non-clonal trees are unique organisms, not genetic duplicates.

Is this a tree or a wildlife hotel?

Trees are vertical eco-habitats, and the older the better, as trunk cracks and intricate branches and leaves give wildlife places to hide, snack and sleep. An old English oak tree (right) can attract an estimated 1,178 invertebrate species, with at least 250 of those depending solely on the oak. It supports 38 species of birds, including redstart, mistlethrush, wood warbler, blue tit and great tit.

Quercus robur

The mighty English oak

Flowers: Either catkins (in dangly clusters) or small pinkish-brown flowers which are eaten by red and grey squirrels and many insects, including the dark crimson underwing moth (*Catocala sponsa*), which relies solely on oaks. Pollen is also popular with bees, including the oak mining bee (*Andrena ferox*), which feeds almost exclusively on oak pollen.

Nuts: The acorn is eaten by 31 different mammals including badger, deer, wild boar, squirrel and wood mouse. Birds include woodpecker, rook, nuthatch and jay.

Leaves: Caterpillars that feed exclusively on oak leaves include the purple hairstreak (*Neozephyrus quercus*), oak lutestring (*Cymatophorina diluta*), great oak beauty (*Hypomecis roboraria*) and merveille du jour (*Griposia aprilina*). New growth attracts aphids, which produce honeydew that wood ants love. (Ants will climb to the top of an oak to track the aphids down.) The spider *Pjilodromus praedatus* lives mainly in oaks.

Bark: Holes and crevices are used as roosts for Bechstein's and barbastelle bats, and birds, including pied flycatcher and marsh tit. The rugged bark also attracts invertebrates such as beetles, butterflies, moths and ants. Larvae of goat moth (*Cossus cossus*) burrow into the trunk. Also beneath the bark are cobweb beetles (*Ctesias serra*) and brown tree ants (*Lasius brunneus*).

Roots and base: An oak base supports 108 species of fungi, above and below ground. These include oak bracket (*Pseudoinonotus dryadeus*) and beefsteak (*Fistulina hepatica*).

Catocala sponsa

Dendrocopos minor

Quercus robur

Lightning and the gods

Oaks are more prone to lightning strikes than most other trees, due to their high water content and the fact they are often the tallest. This is why the tree was associated by the ancient Greeks, Romans, Slavs and Teutonic tribes with their supreme gods Zeus, Jupiter, Dagda, Perun and Thor who ruled thunder and rain. Mistletoe (*Viscum album*) that grows in oak canopies was believed to have been placed there by the gods after a lightning strike.

Purple hairstreak butterfly

Neozephyrus quercus live only in mature oaks (*Quercus robur, petraea, cerris* and *ilex*), with a range throughout much of Caucasia, North Africa and Europe, including Great Britain. The male has a purple sheen to its wings, the female only has patches on her forewings. The eggs look like tiny sea urchins and are laid at the base of oak buds in the summer. They stay there for some eight months, hatching in the spring. Caterpillars feed on buds and young leaves, pupating in bark and leaf matter with butterflies emerging in late June. They feed on aphid honeydew in the tree canopy and can be seen chasing one another, especially in the early evenings of summer.

Meet the dandelion

What's in a name? A lot, actually

This trusty plant goes by many names. The Latin name is *Taraxacum officinale*: the genus name comes from the Arabic word *tarakhshagog*, which means bitter herb, and the species name refers to the medicinal and culinary uses, as the *officina* was the storeroom in a monastery. The name 'dandelion' is from the French *dent de lion*, or lion's tooth: no one really knows why, though it's thought this may refer to its jagged leaf edges. It is more commonly known in France as *pissenlit*, which refers to its diuretic properties. A common name in England used to be 'pee-the-bed', which was adapted from its former name, which Nicholas Culpeper (1616–54) notes in his *Complete Herbal* was 'vulgarly' called *'piss-a-beds'*.

Anatomy of a seed clock

The seedhead of a dandelion has many customs attached to it. Some believe that if you close your eyes, make a wish and blow on a fluff-head, your wish will come true. Others say that this will only happen if all the seedheads are dispersed. The seedheads are also known as fairy clocks: the number of seeds left after you blow will tell you, handily, what the time is in fairyland.

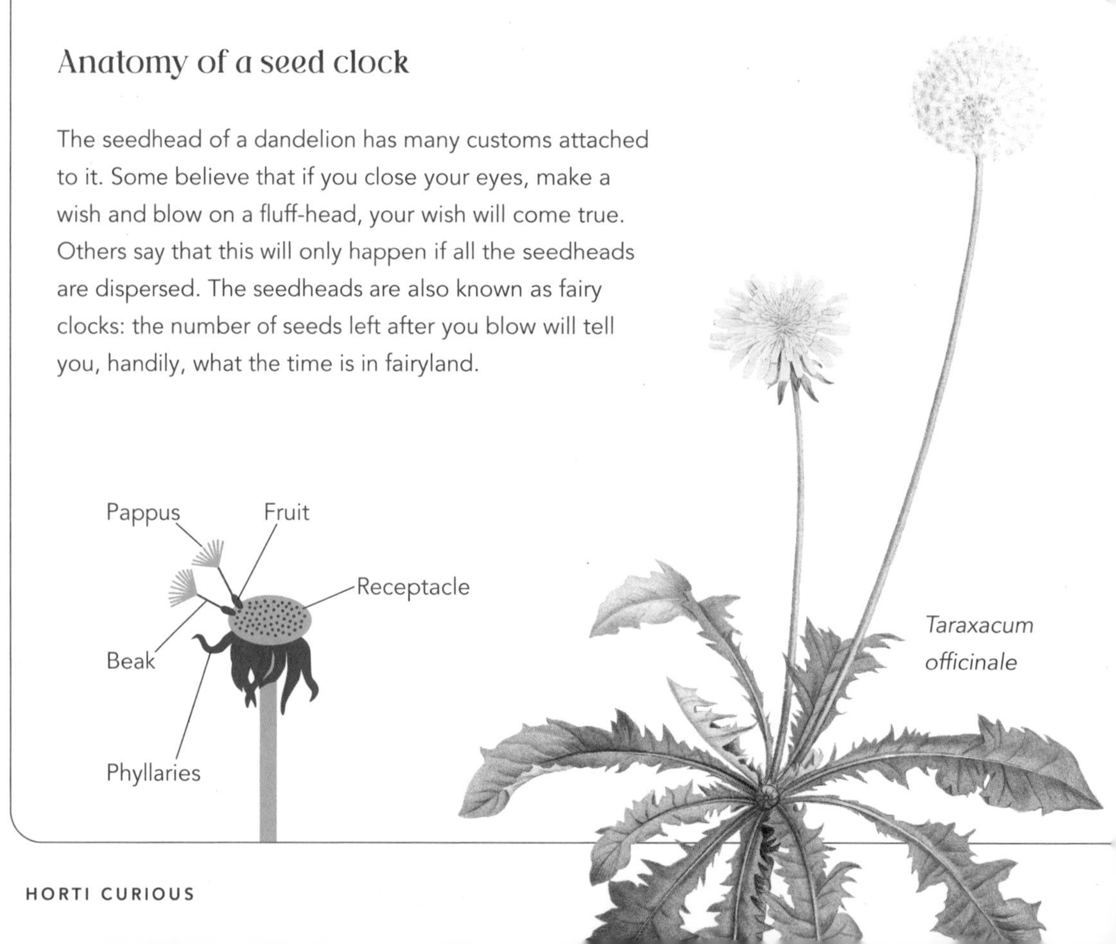

Taraxacum officinale

How to eat or drink a dandelion

Every part of the plant is edible, but even young leaves can be bitter, so soak them overnight in water. Here are a few of its culinary uses:

- Make a tea out of the flowers.
- Eat young leaves as a salad (see the recipe, right).
- Add flower petals to a salad.
- Cook leaves as a vegetable.
- Ferment flowers for wine.
- Use the green plant before it flowers to make beer.
- Grate or chop roots into salads.
- Roast roots as a coffee alternative.

Recipe: Pissenlit aux lardons

- 750g (1lb 10oz) young dandelion leaves
- 1 large shallot
- 300g (10½oz) lardons
- 2 tablespoons wine vinegar
- Salt and pepper
- Optional: small amount of Dijon mustard and oil

1. Clean the leaves and add them and the shallot, peeled and diced, to a salad bowl.

2. Fry the lardons in a pan and, when crisp and dry, tip over the salad.

3. Add the vinegar to the pan and sprinkle over the salad.

4. Add salt and pepper as needed. Mix. A small amount of mustard and oil stirred through the salad is another option.

You can also make this with a mix of dandelion and chicory leaves.

Friend or foe?

Is a dandelion a tiresome weed or an amazing super-plant? The answer is that it is both, and all is in the eye of the beholder. For pollinators in search of nectar it is a godsend, arriving early in spring with an abundance of pollen and nectar, as a result of its flowerheads being composed of hundreds of ray-shaped flowers clustered together. The downside, for lawn neatniks in particular, is that translates into many seeds: a typical puffball seedhead has 150–200 seeds, and there is always more than one flower for every plant. Add in that its long taproot goes deep into the soil, with any small bit able to sprout anew, and it can seem almost indestructible.

Carbon and the climate

The C word

Carbon capture: Trapping carbon dioxide (CO_2) from the air, and storing it, for instance, underground, as a strategy to alleviate climate change.

Carbon cycle: It's natural for carbon to be exchanged constantly among living organisms, soil, rocks, water and air. Humans have distorted the cycle by burning fossil fuels and altering the natural habitat, resulting in a significant increase in CO_2 in the atmosphere. Plants are vital as they absorb CO_2 from the atmosphere through photosynthesis.

Carbon emissions: CO_2 is released into the atmosphere by almost all living organisms through respiration, excretion and decomposition. Humans add to their emissions by burning fossil fuels and through farming and industrial processes.

Carbon farming: Agriculture that aims to enhance and encourage the uptake and storage of CO_2. Examples include avoiding tillage and keeping roots in the ground all year round.

Carbon flux: The amount of carbon moving from one 'pool' to another in a fixed period of time. It cannot be used in isolation to indicate net effects (e.g. whether a practice is beneficial for climate, or carbon balance).

Carbon footprint: The total amount of CO_2 and other greenhouse gases that one person, organization or sector generates. Includes direct and indirect emissions. There are various tools and equations that can help you to measure this for yourself or for any group or organization.

Carbon neutral: When emissions are equal to the amount of carbon being captured and stored.

Carbon offset: A 'credit' that industries or organizations can buy or create to offset carbon emissions. Credits are measured in tonnes of emissions, with one credit per tonne of CO_2.

Carbon pool: General term for all things or locations where carbon can be, which includes the oceans, soil, trees and plants.

Carbon sequestration: The capture, removal and permanent storage of carbon from the atmosphere. One gram of carbon stored for ten years has half the sequestration value of one gram of carbon stored for twenty years.

Carbon sink: Type of carbon pool that is able to take up more carbon from the atmosphere than it releases. Natural carbon sinks include the soil, plants, trees and oceans.

The amazing mangrove

Mangroves (*Rhizophora* spp.) are salt-tolerant, or halophytic, trees that grow in tropical and sub-tropical coastal areas. Their roots form large dense thickets that look like root rafts, creating a buffer zone between the water and shore. They are carbon superheroes, absorbing ten times as much from the atmosphere as land forests. Plus, they are a defence against erosion, improve water quality, mitigate coral bleaching and reduce the impact of storms. Indonesia has the world's largest population of mangroves, covering some 23,000sq km (14,000sq miles), but that number is decreasing due the increase in farmed shrimp production and palm oil plantations.

Coccyzus minor

Rhizophora mangle

Other carbon-tastic trees

Attributes that help carbon storage include: fast-growing, long-lasting, large trunks, dense wood.

- Western yellow pine (*Pinus ponderosa*)
- Hispaniolan pine (*Pinus occidentalis*)
- Douglas fir (*Pseudotsuga menziesii*)
- Bald cypress (*Taxodium distichum*)
- Oak (*Quercus* spp.)
- Horse chestnut (*Aesculus hippocastanum*)
- Blue spruce (*Picea pungens*)
- Teak (*Tectona grandis*)
- Tulip tree (*Liriodendron tulipifera*)
- London plane (*Platanus* × *hispanica*)
- Dogwood (*Cornus* spp.)

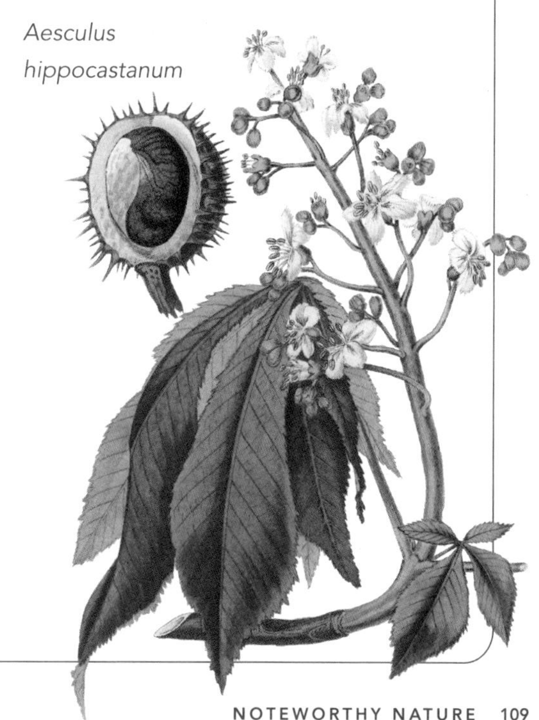

Aesculus hippocastanum

Planting for a low-carbon garden

Trees: Long-lived trees are a great way to store carbon and make a garden climate resilient. Look for drought-tolerant species and/or those that can tolerate waterlogged soils, and make sure that the size of the mature tree fits your space.

Perennials, grasses, weeds: Longer lasting than annuals, with roots staying in the soil.

Hedges and shrubs: Planting a hedge is also a good way to store carbon, and plants with woody stems store more carbon than herbaceous (non-woody) plants.

Fruit and veg: Growing your own means fewer food miles, less buying and more delicious homegrown produce. The most carbon is stored by planting permanent fruit trees and shrubs and perennial vegetables.

Brassica oleracea var. *acephala*

Cynara cardunculus var. *scolymus*

Fragaria × ananassa

Other strategies for a low-carbon garden

Paths without paving: When it comes to hard landscaping, find ways to avoid using concrete and create paths of gravel, or make sure that there are spaces between pavers to minimize water run-off. Water that runs off into public drains and sewers contributes to local flooding.

Power options: Use battery-powered equipment. Avoid petrol or electric mowers/ hedge cutters/leaf blowers unless, for the electric, you have solar panels on the shed roof.

Compost: For peat's sake, it makes sense to create your own compost heap. Avoid buying products that contain peat (read labels carefully); it is an organic material found in wetlands and can store more than twice as much carbon as a forest. When peat is dug up, the carbon is released: at the time of writing this makes up 2–4 per cent of global CO_2 emissions. If you swap your compost for peat free, or make your own, then you will be helping to reduce demand for peat and keeping carbon in the ground.

Natural fertilisers: Use organic fertilisers or make your own out of nettles or comfrey (*Symphytum* spp.).

Use rainwater: Every time you turn on the tap, you are generating carbon emissions. Aiming to always use rainwater stored in a water butt – the bigger the better – will make a significant difference.

Travel miles: Consider how far plants or other items have had to travel to get to you. Wherever possible, buy local.

Symphytum asperum

Chalk and clay meadows

Chalky soil characteristics

- Mostly shallow
- Free draining
- Low fertility
- Alkaline, with a pH above 7

Wildflowers for chalky and limestone soils

- Lady's bedstraw (*Galium verum*)
- Salad burnet (*Sanguisorba minor*)
- Wild carrot (*Daucus carota*)
- Common cowslip (*Primula veris*)
- Ox-eye daisy (*Leucanthemum vulgare*)
- Common knapweed (*Centaurea nigra*)
- Meadow cranesbill (*Geranium pratense*)
- Oregano (*Origanum vulgare*)
- Common toadflax (*Linaria vulgaris*)
- Meadowsweet (*Filipendula ulmaria*)
- Selfheal (*Prunella vulgaris*)
- Field scabious (*Knautia arvensis*)
- Bird's foot trefoil (*Lotus corniculatus*)
- Mountain kidney vetch (*Anthyllis montana*)

Origanum vulgare

Lotus corniculatus

Clay soil characteristics

- Heavy to dig
- Nutrient rich
- Holds water well
- Warms up slowly

Papaver rhoeas

Wildflowers for heavy clay soils

- White campion (*Silene latifolia*)
- Meadow buttercup (*Ranunculus acris*)
- Devil's bit scabious (*Succisa pratensis*)
- Wild carrot (*Daucus carota*)
- Ox-eye daisy (*Leucanthemum vulgare*)
- Ribwort plantain (*Plantago lanceolata*)
- Common cowslip (*Primula veris*)
- Ragged robin (*Lychnis flos-cuculi*)
- Common sorrel (*Rumex acetosa*)
- Tufted vetch (*Vicia cracca*)
- Common yarrow (*Achillea millefolium*)
- Yellow rattle (*Rhinanthus minor*)

Daucus carota

Decoding the rainforest

The forest in layers

Emergent layer: Oversized trees reaching 45–75m (148–246ft) with branches and leaves forming an umbrella over the thick canopy layer below. A busy place for animals and pollinators, including sloths, monkeys, bats, birds and butterflies. Bright sunlight means treetops can get burnt. Usually only a few trees rising above any area. Trees include the kapok (*Ceiba pentandra*) and the Brazil nut tree (*Bertholletia excelsa*), which produces a large fruit that, when it drops, can reach speeds of 50mph (80km/h). Flowers include orchids.

Canopy: This is the main layer and, at 20–30m (66–98ft) up, it is buzzing with high-wire action. This layer uses up 80 per cent of the sunlight and most of the rainwater, with leaves and branches of the dense trees intertwining to form an almost blanket covering. Plants include orchids, mosses, ferns, epiphytes (air plants) and thick climbing vines (lianas) such as philodendron and bush rope (*Strychnos toxifera*).

Understory: Only 2–15 per cent of sunlight makes it to this layer. It forms a dense canopy of smaller trees, shrubs, ferns, climbing plants, bananas. Most plants do not grow beyond 4m (13ft). Relatively few are flowering plants, but the ones that exist tend to be brightly coloured and smelly, to attract pollinators. This is a houseplant bazaar: philodendrons, prayer plants (*Maranta leuconeura*), ferns and zebra plants (*Aphelandra squarrosa*) live in this layer.

Forest floor: Dark (only 2 per cent of sunlight makes it to this layer) and humid with little wind or breeze. Very few larger plants are present; most are mosses, ferns and ginger. Fungi thrive.

Temperate rainforests

Warm in summer, cool in autumn, cold in winter. Annual rainfall 750–1,500mm (29½–59in).

- Appalachian, eastern USA
- Fiordland and Westland, South Island, New Zealand
- Taiheiyo Evergreen Forests, Japan
- Tongass National Forest, Alaska, USA
- Valdivian, southwest Chile
- Atlantic (or Celtic) rainforest, west coast of Scotland and Ireland

Tropical rainforests

Humid, warm and wet. Annual rainfall 2.5–4.2m (8–13¾ft).

- Amazon: Brazil, Bolivia, Colombia, Ecuador, French Guiana, Guyana, Peru, Suriname and Venezuela
- Bosawas Biosphere Reserve: Nicaragua
- Congolian rainforest: Democratic Republic of the Congo, Republic of the Congo, Cameroon, Equatorial Guinea, Gabon, Central African Republic
- Daintree: Queensland, Australia
- Kinabalu National Park: Malaysian Borneo
- Monteverde Cloud Forest: Costa Rica
- Rainforests across Southeast Asia, Indonesia, Borneo and Sumatra

The 'great flower'

Victoria amazonica is a waterlily that has huge leaves with upturned fluted edges, like giant green tart tins, which float in the shallow waters of the Amazon Basin. In Brazil, they are called *uape jacana* ('the lily trotters water lily') and in the Quechuan language, *atun sisac* ('great flower'). The actual flowers float among the huge leaves, and last from just 48 hours to three days, turning pink once their pollen has been released. They are pollinated by a type of scarab beetle which is actually trapped inside the flower overnight. The leaves are anchored by stalks that can reach up to 8m (26ft) long.

Can murderers also be saviours?

Strangler figs, which take in several species from the genus *Ficus*, can be seen as both friend and foe of the rainforest. Their sticky seeds are deposited in the canopy layer in animal droppings. They grow up and out (the plentiful fig foliage reaching across the canopy) and down the trunk of their host tree, forming an intricate latticework that can be suffocating. When the roots reach the soil, they tend to grow out, surrounding the host tree roots. It's no mystery why stranglers have the Spanish nickname *matapalo*, or tree-killer. But they can also, in some circumstances, be saviours: they can stabilize trees, protecting them from being felled during storms. It's also a keystone species and provides a home for bats and birds.

Heliconia densiflora

Lobsters of the forest

The common names for various species of *Heliconia* say it all in the most colourful way: they are known variously as hanging lobster claw, toucan beak or false bird-of-paradise. They range in size from 0.5–5m (1½–16½ft), with large leaves and what look like bright flowers but are actually bracts. These are sometimes erect, other times drooping, coloured in shades of scarlet, orange and yellow. Most are pollinated by hummingbirds that rely on the plants for both food and nesting.

Heliconia psittacorum

Under the deep blue sea

There's really only one ocean plant

There are five types of plant organisms in the ocean – phytoplankton, algae, kelp, sargassum and seagrass – but only one is a flowering plant and that is seagrass. It is the world's only flowering plant that can live under sea water. Its leaves tend to be thin but very flexible, as they must constantly 'go with the flow' of waves and currents. Seagrass disperses its buoyant seeds and pollen using the tides and currents, the equivalent of wind on land. Most grow in coastal waters 1–3m (3–10ft) deep, but one, a paddle grass (*Halophila decipiens*), has been found 58m (190ft) deep. Imagine making oxygen all that way down – and without a mask.

Thalassodendron ciliatum

Callophyllis ornata

Sargassum

Underwater meadows

Seagrass meadows are beautiful flowing masses of leaves that provide incredible value in terms of their ecosystem. One hectare (2½ acres) of seagrass meadow can provide a home for 80,000 fish and 100 million small invertebrates. They are nurseries for all the small fry and are fantastically efficient at capturing carbon. On a clear day, underwater, you can see the oxygen bubbles forming and heading to the surface. These meadows are carbon sinks (see page 108) that act as giant sponges, transferring carbon to the ocean floor 35 times more effectively than tropical rainforest systems transfer carbon to the soil. This makes them planet saviours but they are under threat from pollution, algal blooms, increasing water temperatures and storms.

The dugong depends on it

A perfect example of the power of seagrass can be found on the Great Barrier Reef, where 15 species live, covering 4.5 million hectares (11 million acres). It's the largest seagrass ecosystem on the planet. The plants provide food for the reef's largest grazers – the turtle and dugong, which eat, respectively, 2 and 30kg (4½ and 66lb) of seagrass a day. Significant populations of the endangered dugongs (also known as sea cows) live in the reef; they would not survive in these waters without seagrass meadows. The plant is seen as an ecosystem engineer, acting as a nutrient and sediment filter, filtering pollution to 'clean' the water before it reaches the coral areas. Seagrass meadows are often called the 'lungs of the ocean', but in this way they also serve as the kidneys.

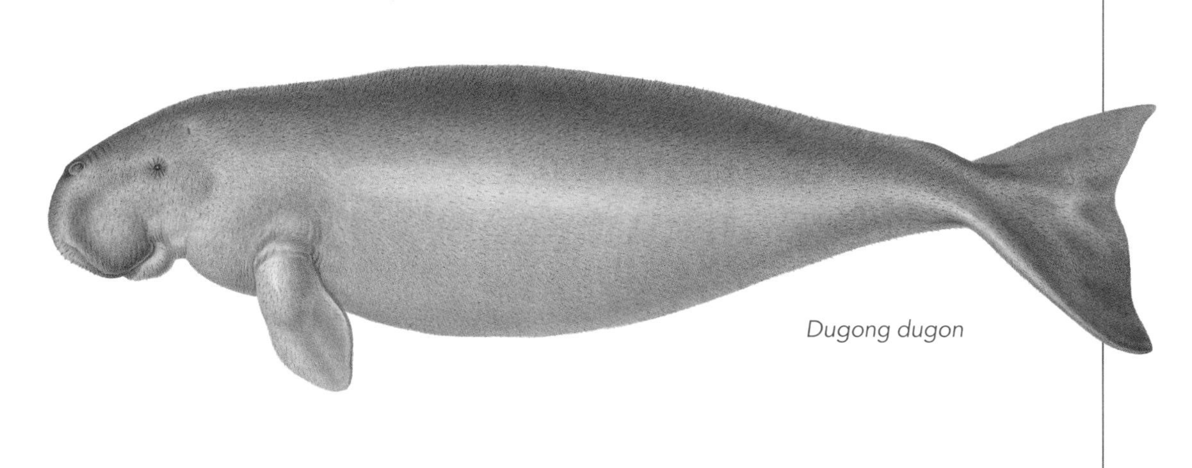

Dugong dugon

Going batty

Bat facts

- There are more than 1,300 species of bat in the world.
- Bats live in all parts of the world except the Arctic, Antarctica and some oceanic islands.
- The largest is the flying fox (*Pteropus*), with wingspans of 1.5m (5ft) and more, in Asia, eastern Africa and Australia.
- The smallest is the bumblebee bat (*Craseonycteris thonglongyai*), weighing 2g (0.07oz), found in Thailand and Myanmar.
- Most bats feed on flying insects but some species eat frogs, fruit, nectar, pollen, fish and even other bats. Plus, of course, blood, ergo the vampire bats (*Desmodontinae*).
- Bats are the only mammal that can truly fly.

Pteropus poliocephalus

Helpful bat practices

- Some bats are pollinators, flying from flower to flower.
- They eat thousands of insects a night.
- Some eat fruit and, through their droppings, spread the seeds. Fruit-eating bats account for almost all seed dispersal for early growth in cleared rainforests.

Invite bats to your garden

- Plant night-scented flowers to attract insects for insect-eating bats.
- Create a pond.
- Plant linear hedges/treelines.
- Put up bat boxes.
- Reduce artificial lighting.

A year in the life of a bat

Winter: Bats are hibernating, using stored fats from summer and autumn feeding as fuel. They may roost singly or in small groups, often in old trees, caves and derelict buildings.

Spring: Their fat reserves run low in early spring and, on warmer nights, they may succumb to their hunger, heading out to find insects and water. Migratory bats will move on now. In late spring, females form maternal colonies and search for nursery sites (usually trees or buildings). Males roost on their own or in small groups.

Summer: Females give birth to a single pup (or sometimes more than one), which they feed with their milk. Adult bats now catch thousands of insects a night. At three weeks the offspring start to learn to fly, and by six weeks they can catch insects. Maternal colonies disperse.

Autumn: This is the mating season, with males using special calls including purrs, clicks and buzzes to attract females. Autumn is also the season for migration or hibernation. They start to build up fat reserves before winter, and in mid- to late autumn begin periods of torpor, which then become longer.

Wasting away in margarita-land

In Mexico, bats have had a long relationship with blue agave (*Agave tequilana*), which is the plant with large spiky leaves that gives us tequila. Three species of bat, all with very long tongues, feed on agave nectar, which they do at night, also pollinating them as they fly from plant to plant. It is the core, or *pina*, of this succulent that is juiced and fermented to make tequila. A worldwide boom in the stuff that makes margarita-land go round has resulted in many farmers in Mexico using cloned plants and harvesting before flowering. This has been bad for the bats (lack of food) and the plants (lack of genetic diversity leads to diseases), and has led to a campaign for people to buy bat-friendly tequila. The pollination of plants by bats is called chiropterophily. Don't try to say that after a tequila shot or three.

Life at the edges

On the edge

Edges, such as shorelines, riverbanks and forest paths, are where landscapes change. They can be natural or created by us. On edges, a landscape is neither one thing or another, but a place of transition that attracts a mix of wildlife from each landscape and maybe a few others besides.

Definition

Ecotone: *noun*. A zone of transition between two biological communities. From ECO(LOGY) + –tone, from Greek *tonos*, for 'tension'.

Manmade edges

You and I may not want to live by the side of a road but, for a plant, it has its upsides. Roadside verges are mostly undisturbed and the soil is nutrient poor, which is preferred by wild plants and flowers. The seeds of wildflowers and weeds arrive, blown in by nature. Shrubs and trees are planted by both humans and nature (birds and squirrels). Even rubbish can be beautiful once that apple core becomes a tree. In the UK, verges are home to more than 700 wildflower species, including 29 of Britain's 52 wild orchids. In many countries these verges are managed as conservation areas. Other names for these corridors include 'nature strip' and 'sidewalk lawn', or the excellent description of 'besidewalk'.

Natural fringes

- Marshlands, between dry and wet
- Mangrove forests, between terrestrial and marine
- Grasslands, between desert and forest
- Estuaries, between saltwater and freshwater
- Riverbanks, between flowing water and dry land
- Tropical edges, between rainforest and crop or grassland

Maximize garden edges

Build a dead hedge and the beetles and insects will come to stay, and then the mice will see it as a cosy home (rather than inside your house), and then put up an owl box, so the owls will come to hunt the mice. From no wildlife to a chain with an apex predator in less than a year.

A dead hedge can be as small as 1.5m (5ft), the length of one fence panel, or as long as you want. Save twigs and branches after pruning or after a storm. The hedge can be used as a boundary or zone divider.

You will need:

- Branches or garden stakes
- Mallet
- Dead plant material
- Flowering climbers

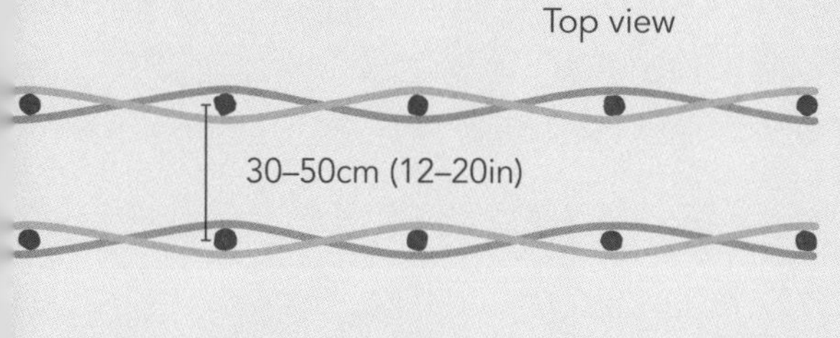

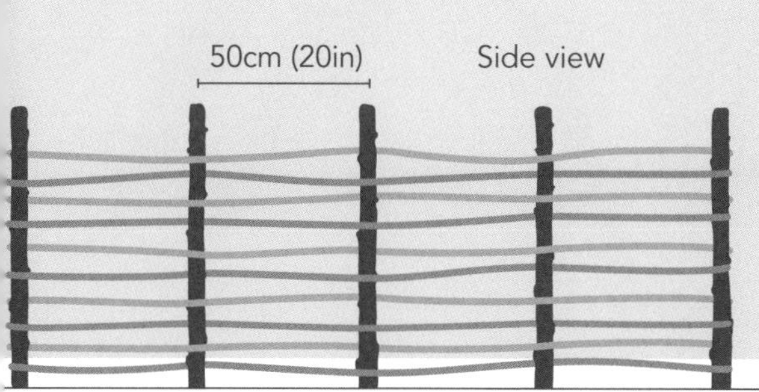

1. Select branches 3–5cm (1–2in) in diameter. They can be almost any height up to 2m (6½ft). If you don't have dead material to hand, garden stakes can be used. One edge needs to be cut at an angle. (Tip: Don't use willow unless you want a hedge that sprouts.)

2. You are making two parallel rows of stakes, approximately 30–50cm (12–20in) apart.

3. Space your stakes about 50cm (20in) apart in your parallel rows.

4. Pound them into place with a mallet.

5. Use dead material from pruning or a storm – cut up any big branches, a discarded Christmas tree etc. Weave branches and twigs in and out of the stakes or pile them up in the middle.

6. Add colour in the form of flowering climbers.

7. Add wildlife interest such as a hedgehog box.

8. Top up whenever you want: the more you weave in the twigs the better the habitat for birds' nests and other wildlife.

Kentiopsis macrocarpa

Chapter 4
Amazing Spaces

Here is a look at horticultural place-making in all its varied forms: large and small, organized and guerrilla, real and fantasy, urban and wild. Plants can shape-shift the way we live: with parks in the sky, gardens in the Arctic and follies in our gardens.

Treasured parks

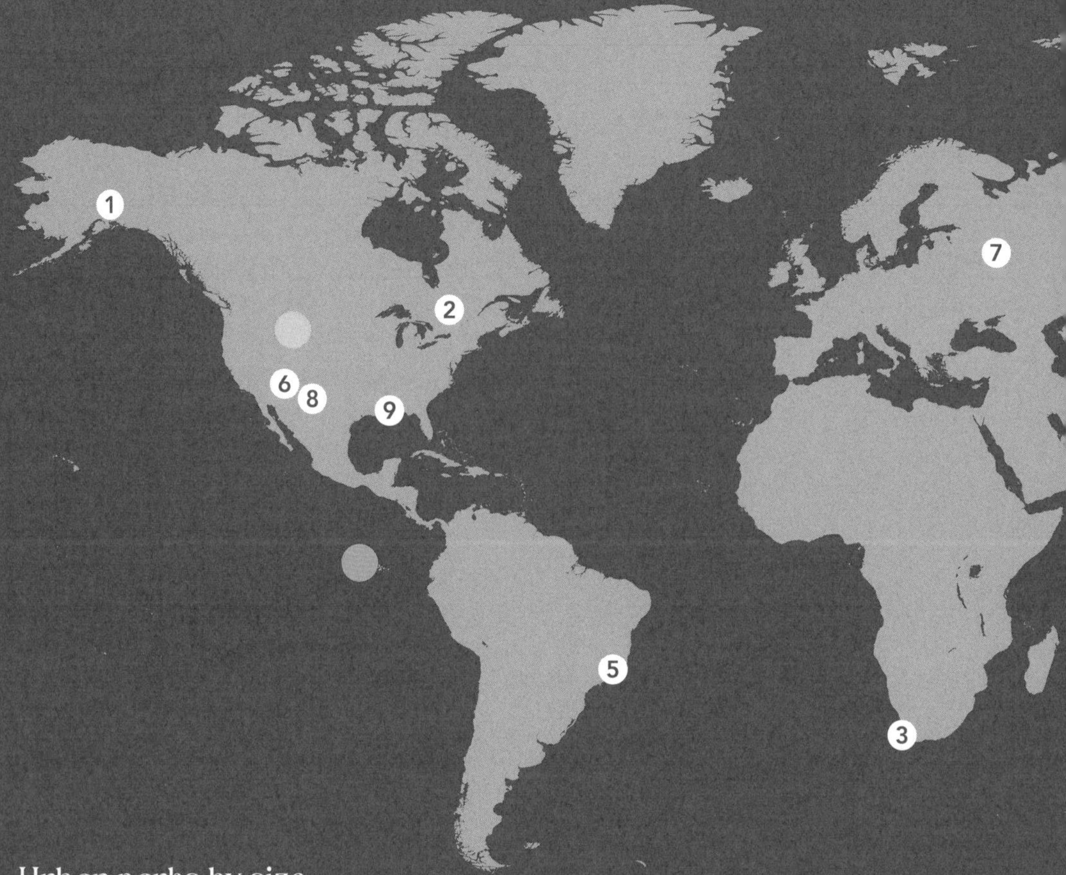

Urban parks by size

1. **Chugach State Park**, Anchorage, Alaska, USA: 200,400 hectares (495,199 acres)

2. **Gatineau**, Quebec, Canada: 36,700 hectares (89,205 acres)

3. **Table Mountain National Park**, Cape Town, South Africa: 22,100 hectares (54,610 acres)

4. **Margalla Hills National Park**, Islamabad, Pakistan: 17,386 hectares (42,961 acres)

5. **Pedra Branca State Park**, Rio de Janeiro, Brazil: 12,400 hectares (30,626 acres)

6. **McDowell Sonoran Preserve**, Scottsdale, Arizona, USA: 12,300 hectares (30,394 acres)

7. **Losiny Ostrov National Park**, Moscow, Russia: 11,600 hectares (28,664 acres)

8. **Franklin Mountains State Park**, El Paso, Texas, USA: 9,810 hectares (24,246 acres)

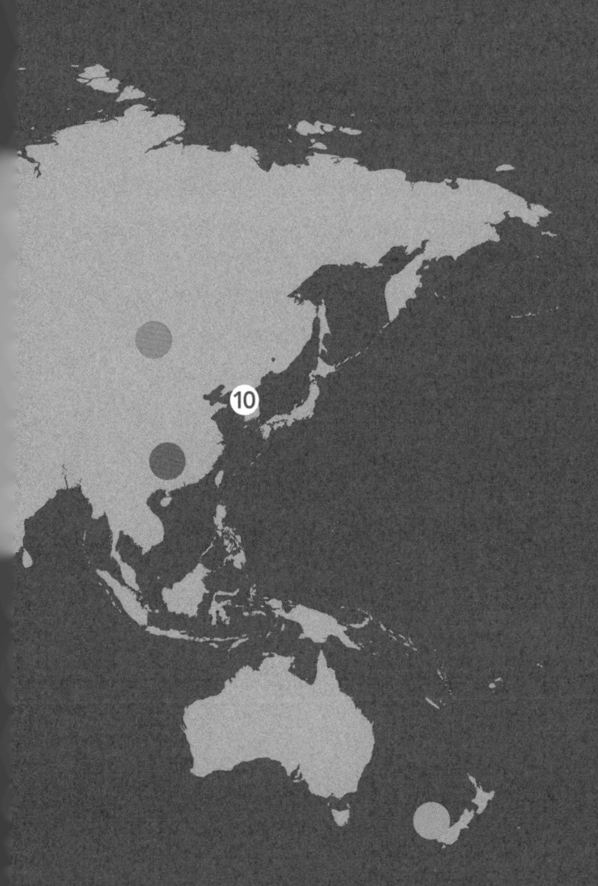

9. **Bayou Sauvage Urban National Wildlife Refuge**, New Orleans, Louisiana, USA: 9,210 hectares (22,758 acres)

10. **Bukhansan National Park**, Seoul, South Korea: 7,992 hectares (19,748 acres)

Incredible parks of the world

- **Yellowstone**, USA: Founded 1872, the first national park in USA. 8,991sq km (5,587sq miles), mostly in Wyoming. Geysers, hot springs, canyons, waterfalls. Wildlife includes buffalo, grizzly bears and wolves.

- **Bogd Khan Uul**, Mongolia: First protected in 1783, in the Khentii Mountains. 673sq km (419sq miles), south of Ulaanbaatar. Habitats include alpine tundra, grassland, taiga and snow forest.

- **Galapagos**, Ecuador: Founded 1959, located 800km (500 miles) west of mainland Ecuador. The park covers 7,880sq km (4,896sq miles), with islands renowned for their specific flora and fauna.

- **Guilin Lijiang**, China: Northeastern region of Guangxi Zhuang, dramatic landscape of limestone karsts and vast bamboo forests. The river flows a total of 164km (102 miles).

- **Fiordland**, New Zealand: Founded 1952, South Island. 12,697sq km (7,890sq miles). Dramatic mountainous landscape, glaciers and fiords.

The largest . . .

The Northeast Greenland National Park is 972,000sq km (375,000sq miles) and has a permanent population of polar bears and muskoxen, but no people. To imagine the park size, think of Spain and France put together. This is a High Arctic wilderness with stunning scenery and vast landscapes, fjords, mountains and icebergs. Eighty per cent of the area is covered by the Greenland Ice Sheet. There are no roads, harbours, hotels, guesthouses or commercial airports. There are a few meteorological, research and military stations, including the elite naval unit Sirius Dog Sled Patrol, and they are accessed by private gravel airstrips. No hunting or fishing is permitted except by those who live in the local community of Ittoqqortoormiit, south of the park, though they have not done so for many years. No motorized vehicles or all-terrain motorcycles are allowed. You can camp in summer, but no one is permitted to take anything out of the park. The result is that the High Arctic habitat for plants survives and thrives – with jewel-coloured alpines such as purple saxifrage (*Saxifraga oppositifolia*) as well as the subtler yellow of the Icelandic poppy (*Papaver nudicaule*) and the waxy Arctic bearberry (*Arctostaphylos uva-ursi*), which provides a (very small) polar bear snack.

Yellowstone National Park

Arctostaphylos uva-ursi

. . . And the smallest

It was 1946 when a man named Dick Fagan returned from the Second World War to his job as a journalist at the *Oregon Journal*. The view outside his office window, in downtown Portland, was not particularly inspiring, being an empty hole in a central reservation where a lamppost used to be placed. Fagan decided to beautify the hole, planting flowers. He wrote a whimsical column called 'Mill Ends' (a reference to the irregular bits of leftover lumber at mills) and often referred to the now blooming hole, which he called the World's Smallest Park. He described various events, which seemed mostly to involve a group of leprechauns. Mill Ends officially became a city park on St Patrick's Day in 1976 and, over the years, it has had many donations, including a small swimming pool with a diving board for butterflies. It is 61cm (2ft) across with a total area of 2,916sq cm (452sq in, or 0.00007205784 acres).

Paradise gardens

What paradise means

The word 'paradise' was originally derived from a word in Old Avestan, one of the Iranian languages, which was *pairidaēza* and means 'an enclosure'. In ancient Persian, the word applied to a garden – a green space that was walled, protected and separate from the harsh world of the desert. This idea, that a garden is a place of respite from a surrounding world, somewhere that is profoundly different, has lasted through the centuries. The first 'paradise garden' is from the reign of Cyrus the Great, who founded the Achaemenid Empire in 550 BCE. Now, it is sometimes also called an 'Islamic garden' (see page 46), having been adopted through the spread of the Muslim Arabic conquests as far east as India and as west as Spain.

Paradisea liliastrum

'Look at the water and look at the basin, and you will not be able to tell if it is the water that is motionless or the marble which ripples.'

Translation of the inscription on a fountain in the Alhambra, Spain

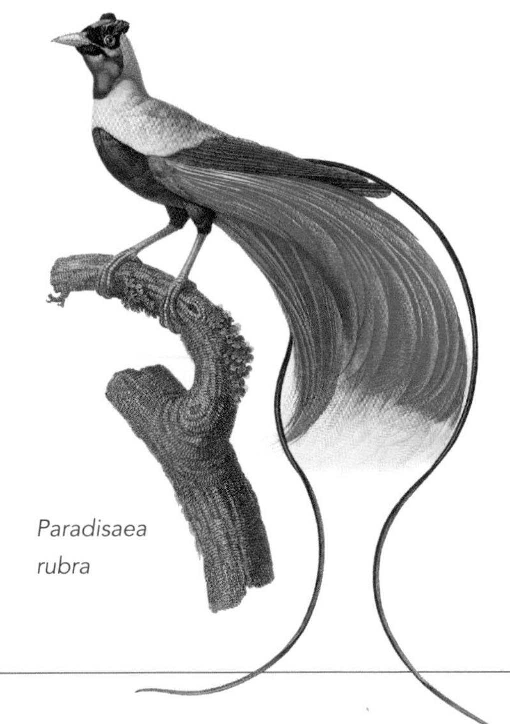

Paradisaea rubra

Blueprint for paradise

FORMAL. CALM. SYMMETRICAL. ENCLOSED.

Geometry is everything. The classic layout is four squares with a central pool that is usually rectangular. Often, each of those four squares is then divided into four more squares. This is called *charbagh*, 'four gardens'.

Water is the second essential aspect of a paradise garden. The central pool, which is sometimes circular, usually has four 'rills' that extend in each direction, separating the squares. These are said to signify the four rivers in Allah's gardens of paradise: the rivers of water, wine, milk and honey.

Punica granatum

Paradise gardens are inward looking, and the surrounding walls create a sense of privacy, a barrier that separates a private world from that beyond. The walls also keep out unwanted plants and provide a precious commodity in desert life – shade.

Fruit trees are essential to the spirit of this type of garden. Orange trees, with their beautiful fragrant flowers, are often linked to them, as are olive, date, pomegranate, cherry, apple, fig and almond.

Scent is important to the planting. Roses and lilies, as well as narcissi in spring, will be part of the planting. Traditionally, the four squares will either be slightly sunken or the path between them raised, so that as you walk, the scents of the garden surround you.

Sound and movement of water are valued, as long as they are calming and meditative, such as a tinkle of water as it drops down a step, or a gentle rill ripple.

The overall philosophy is 'less is more' and the idea is that the sum is greater than its parts. The garden aims to be contemplative and quiet, meditative and spiritual, allowing for reflection and an appreciation of all that this entails.

Lost in mazes

Labyrinths and hedges

The maze has an ancient history, spanning thousands of years and mentioned in Greek myth and by Roman poet Virgil. We now use the terms labyrinths and mazes interchangeably, though the former is unicursal, the latter more complicated, with intricate loops and blind walks. Labyrinths came first, circular puzzles that made ancient Romans and Greeks go round in circles as they attempted to reach the centre. These were not hedges but patterns on the ground, to be seen as well as walked. The pagan puzzle became, with Christianity, a religious journey of the soul. The idea of finding the right path of the labyrinth was now seen as a spiritual process, something to meditate on as you looked for clues to reach the centre (salvation).

Hedge mazes were much more playful and grew up, literally, from the knot gardens of Renaissance Europe. These were intricate creations, epitomized by the labyrinth of Versailles, France, a hedge maze that included fountains and sculptures depicting Aesop's fables. The maze itself was created in 1668 by landscape architect André Le Nôtre and, a few years later, the sculpted animals were added at intersections, the jets of water from their mouths designed to look as if they were talking to one another. Sadly, Louis XVI had this fantastical creation removed in 1775, replaced by an English garden in homage to his queen, Marie Antoinette.

Hedging plants

The only surviving example of a Baroque hedge maze is at Hampton Court, London, UK, commissioned by William III *c.* 1690: it is a puzzle maze, trapezoid in shape, now made of yew though originally hornbeam. Today, hedge mazes are often planted as tourist attractions. Box blight means that box is not used as much compared to these traditional hedging plants:

- Yew (*Taxus baccata*), evergreen
- Holly (*Ilex*), evergreen
- Hornbeam (*Carpinus*), holds some leaves during winter
- Beech (*Fagus*), holds some leaves during winter

The Peace Maze
Castlewellan, County Down,
Northern Ireland, UK. 2001

Marlborough Hedge Maze
Blenheim Palace, Oxfordshire, England, UK. 1987

Leeds Castle
Broomfield, Kent, England, UK. 1987

Hever Yew Maze
Hever Castle, Edenbridge, Kent, England, UK. 1906

Villa Pisani
Stra, Venice, Italy. *c.* 1720

Traquair House Maze
Innerleithen, Scotland, UK. 1981

Hampton Court Maze
Hampton Court Palace, East Molesey,
England, UK. *c.* 1690
The oldest hedge maze in Britain.
On average, it takes 20 minutes to reach the centre.

Guerrilla gardening

The power of green

Guerrilla gardening is all about transforming urban streets through radical acts of gardening. The name itself came out of the Green Guerrilla movement in New York City in the 1970s, when a group of people started throwing seed bombs (also called 'green aids') over the fences of abandoned lots and planted sunflowers in the centre strip of the city streets. Suddenly abandoned buildings started to sprout window boxes. When the Green Guerrillas looked at a large, abandoned lot, littered with rubbish and debris, at the corner of Bowery and Houston Streets, they saw green, not red. They began to grow, creating the Bowery Houston Community Farm and Garden, which still exists today (though renamed as the Liz Christy Bowery Houston Garden), a vibrant green space with magnolia and weeping birch, ponds, vegetables, herbs and flowers. Twenty volunteers keep it going, and in 2023 the Green Guerrillas celebrated their 50th anniversary.

'You put the seeds in your hand and you toss them. It's that simple. It's just a beautiful, accessible thing for almost anyone.'

Phoenix McGee of 'SF in Bloom', TikTok

Magnolia × soulangeana

Coleslaw, anyone?

Is it art or is it guerrilla gardening? 'The Vacant Lot of Cabbages' in the centre of Wellington in New Zealand managed to be both. In 1978, a group of artists planted 180 cabbages on an abandoned lot in the middle of the capital in such a way that it spelt out the word 'cabbage'. The installation lasted for six months, with many residents getting involved as the brassicas grew larger. It culminated in a week-long harvest festival with poetry readings, performances and the distribution of free coleslaw. Photographs of the cabbage patch are now in the National Library and the Museum of New Zealand.

Brassica oleracea var. *rubra* (Capitata group)

Essential guerrilla kit

- Seed bombs
- Parmesan cheese shaker (for seed)
- Trowel
- Watering can/bottle
- Bag to carry away rubbish
- Sturdy gloves for rubbish removal and planting

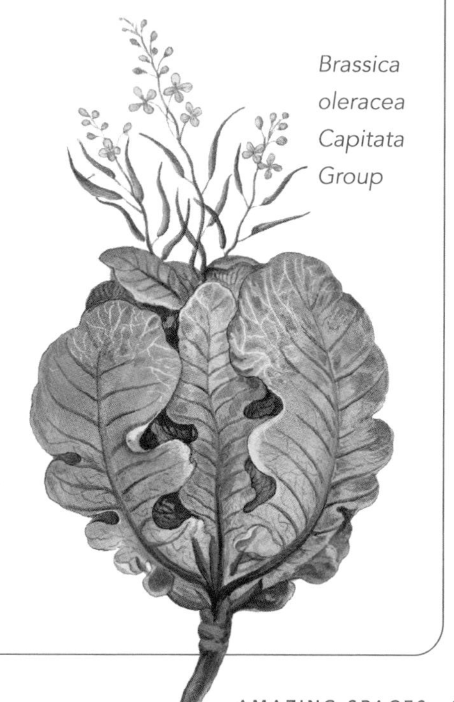

Brassica oleracea Capitata Group

Green guerrillas of the globe

- **Organic Starters** transformed an empty piece of land in Nørrebro, Copenhagen, Denmark, in 1996 into a garden in one night.
- **Green guerrillas** in Poland started running the Shelter for Unwanted Plants in 2010, and in 2017 established the World Day of Planting Pumpkins in Public Places (16 May). In 2020 the same group launched a 'National Suspension of Lawn Mowers' campaign.
- **Wild Zone** in Finland has a wide variety of activities including introducing guerrilla meadows to cities and towns.
- **Greenaid**, in Los Angeles, USA, has placed old gumball vending machines in at least 20 outposts around the city. They dispense seedballs that have been prepared for the specific area.

Make a seed bomb

Seed bombs are a key guerrilla gardening tool (though they can be used in any garden). Spring and autumn are the best time to bomb, as they tend to be wetter and rain is needed to break up the bomb. The seed bombs should be small – think ping-pong ball – and there is no need to fret about a correct throwing technique: simply lob and go (and grow). Aim them at bare patches of soil in your own garden or, if you are on manoeuvres (or simply walking somewhere), look for neglected street planters, tree beds, roundabouts or abandoned plots.

Ingredients

- Native meadow flower seeds (to avoid invasive species)
- Peat-free, organic compost
- Powdered clay (from craft shops)
- Chilli powder (optional)
- Water

1. Mix together 1 part seeds, 3 parts clay powder, 5 parts compost.
2. Add a pinch of chilli powder to deter squirrels from eating your bomb.
3. Slowly mix in water with your hands until the mixture sticks together, but it should not be too wet.
4. Roll the mixture into a small firm ball. (Remember: the smaller the ball, the easier to throw.)
5. Put the ball(s) in a cardboard box in a warm, dry place to dry out.

Penstemon parryi

Botanic gardens

Botanical beginnings

From the very start, botanic gardens were always about something other than just plants. The driving force of the first such gardens, which date from the 16th century, was iatrochemistry, the academic study of medicinal plants. In the Age of Enlightenment and into the next centuries, botanical gardens became inextricably linked with colonialism, with plant hunters scouring the globe to return with 'new' species for research, study, categorization and display. Latterly, botanic gardens have embraced two separate purposes. The first is conservation and studying plants in the face of climate change. The second is tourism, with many now including various wow-factor attractions to appeal to visitors of all ages. What do all botanic gardens have in common through the ages, other than plants? A belief in the power of science and, crucially, plant labels.

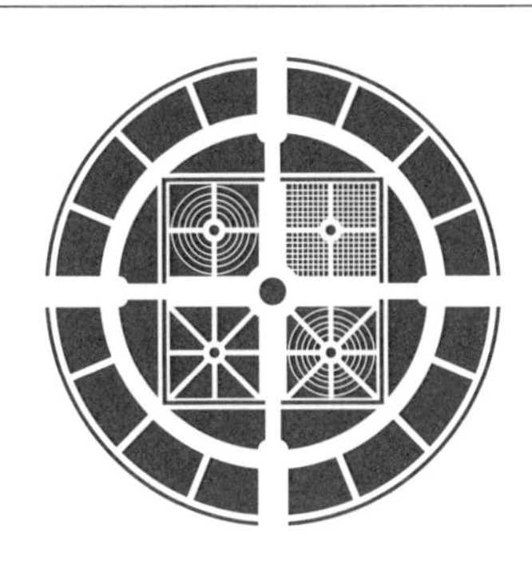

Back in time

The Orto Botanico di Padova in Padua, Italy, founded in 1545, is a botanical time-travelling portal. This is the first botanic garden, on land that once belonged to a Benedictine monastery, and its original layout is intact. When you think how people like to chop and change everything, and especially gardens, this makes it very special. The layout consists of an outer circle (84m/276ft diameter) enclosing a square split into four quarters by intersecting paths. This symbolized the world as they knew it, with the entire garden surrounded by water (oceans). The four quarters were arranged in geometrical patterns (see above), apparently to help students remember what was planted where and for what purpose. It was originally for medicinal plants (a high wall had to be built in 1552 as night thefts of the valuable 'cures', or plants, were common) but the collection soon expanded with additions from Venetian traders. The oldest plant in the garden is from 1585, known as Goethe's Palm.

Meconopsis simplicifolia

Don't forget your skis

Tromsø's Arctic–Alpine Botanic Garden, in Norway, opened in 1994 and is unusual in more ways than one. It is inside the Arctic Circle, which makes it the furthest north of any botanic garden, and it can be visited 24 hours a day, with no gates or fees. Plants from arctic and alpine regions grow in naturalistic settings of huge boulders and various rockscapes. You can visit on skis during the winter, and even though it is permanently dark in December and the first half of January, the Aurora Borealis (Northern Lights) provides a different type of backdrop. A plant highlight is the Sikkim poppy (*Meconopsis simplicifolia*).

Other best botanicals

Dramatic setting: Kirstenbosch National Botanical Garden, South Africa, covering 520 hectares (1,285 acres), featuring mostly native South African plants, set against the grandeur of the eastern slopes of Cape Town's Table Mountain.

Blooming sandscape: Desert Botanical Garden in Phoenix, Arizona, USA, 53 hectares (130 acres), has a huge collection of plants adapted to desert conditions including 4,026 agaves and 14,000 cacti, and features ecosystems including a mesquite bosque, upland chaparral and semi-desert grassland.

Historic and comprehensive: Royal Botanic Gardens, Kew in London, UK, founded in 1759, has the largest and most diverse collection of plants in the world on a 120-hectare (300-acre) site.

Tree-mendous: Acharya Jagadish Chandra Bose Indian Botanic Garden in Howrah, West Bengal, India, 45 hectares (110 acres), founded in 1787. Its outstanding sight is The Great Banyan Tree, more than 250 years old, covering 1.5 hectares (3¾ acres).

Greenhouse shapes

Palm House, Royal Botanic Gardens, Kew, London, England, UK. 1844

Grosses Palmenhaus Schönbrunn, Vienna, Austria. 1882

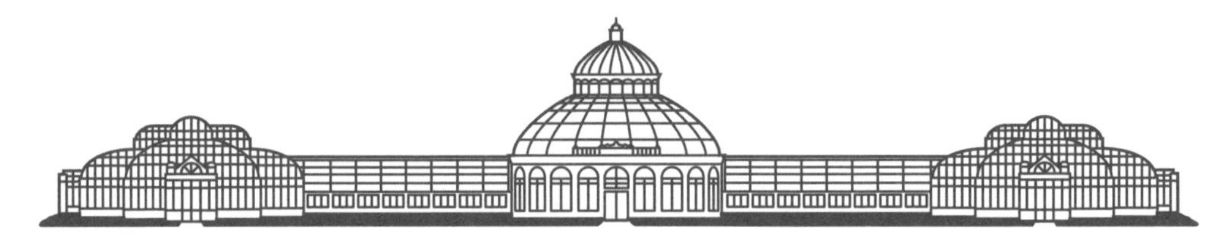

Enid A. Haupt Conservatory, New York City, USA. 1902

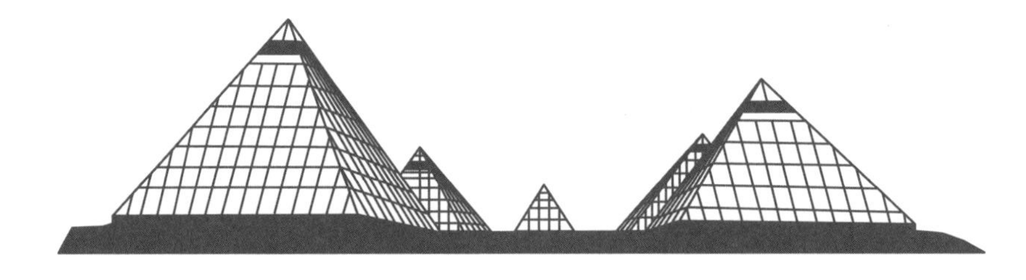

Muttart Conservatory, Edmonton, Canada. 1976

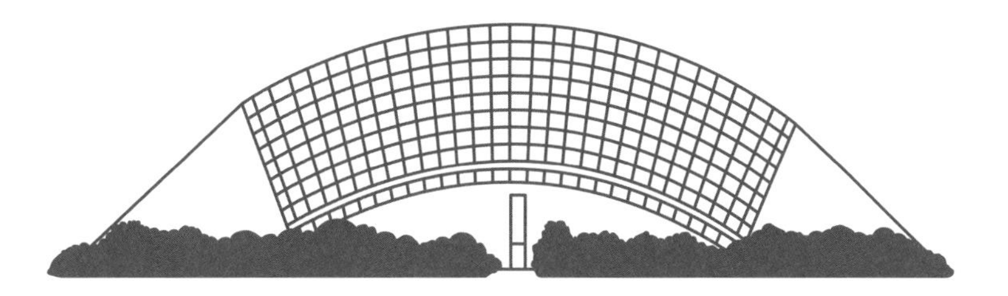

Bicentennial Conservatory, Adelaide Botanic Garden, Australia. 1988

Greenhouse, Botanical Garden of Curitiba, Brazil. 1991

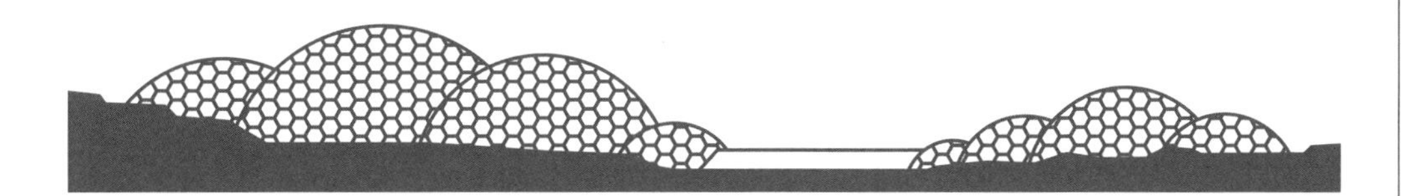

Geodesic biome domes, Eden Project, Cornwall, England, UK. 2001

Flower Dome, Singapore, 2011

Tree life

What lives in trees?

Nests are often hidden spaces in a garden, nestling inside dense thickets of climbers such as ivy, and in hedges and trees. Each bird creates a different nest structure and most of their materials come from the garden – lichen, moss, grasses, skeleton leaves, fresh leaves and so on.

Collared dove (*Streptopelia decaocto*)
These nests look ramshackle and are, essentially, a bunch of twigs thrown together, with no lining. It's a miracle the eggs don't just fall through and break. The nests are built in trees or climbing vines from 2–20m (7–66ft) up. The doves usually have three to five broods a year.

Wren (*Troglodytidae*)
The males make between five and eight nests using small twigs, grasses, moss and dead leaves. Once the female has chosen which nest she wants, she will line it with feathers and hair. The nests are often hidden in ivy against a wall.

Brolga (*Grus rubicunda*)
Brolgas, also known as Australian cranes, do not migrate and have been known to use the same nesting site for up to 20 years. Their nests, built by both sexes, are in shallow marshes, made of uprooted grass and other plant material, and form a mound for a small island or are, occasionally, floating.

Baya weaver (*Ploceus philippinus*)
This bird uses grasses, palm fronds and other leaves to create intricate hollow teardrop-shaped nests that hang from branches of thorny trees or palms. To thwart any unwanted visitors, the final touch is a downward-facing entry tunnel.

Social weaver (*Philetairus socius*)
These birds create large intricate nests that are divided into nesting chambers. There can be up to a hundred chambers in one single nest. They use large twigs to form the roof, and grasses to create the various rooms, which have varying temperatures as per their location (inner or outer). Preferred trees include the camel thorn (*Vachellia erioloba*) and shepherd's tree (*Boscia albitrunca*).

Harris's hawk (*Parabuteo unicinctus*)
These South American birds of prey are adapted to dry habitats and build their nests in the treelike saguaro cactus (*Carnegiea gigantea*) or mesquite, or palm or pine trees. The large bulky structures, up to 60cm (2ft) wide, comprise sticks and parts of cactus, and are lined with bits of cactus, grass and feathers.

Ploceus philippinus

Other tree lovers

Key requirements: It helps if you can fly or have a good tail to balance. Other helpful attributes are opposable thumbs, no fear of heights and the ability to sleep almost anywhere.

Koalas: Mostly live in eucalyptus; don't make nests but sleep in tree forks.

Sloths: Preferred tree is the cecropia (also known as, yes, the sloth tree); sleep hanging upside down.

Temperate bats: Cavities in deciduous trees such as oak and beech; hang upside down to sleep.

Treefrogs: Live in a variety of trees; some make 'foam' nests, others lay eggs on leaves.

Tree kangaroos: Marsupials adapted for life in rainforests; they sleep sitting on a branch.

Aye-ayes: Nocturnal lemurs that live only in Madagascar; make ball nests for daytime sleeping.

Humans: Manmade treehouses are more visited than lived in, for recreation, work or just hanging out, in this case literally. Only the Korowai, also called Kolufo, people of south-eastern Papua New Guinea, are thought to live in tree houses deep in the rainforest.

Superbloms

Super-special

It's a rare thing for a desert to burst into bloom, covering not just one hill or one valley with colourful wildflowers but an entire mountain and, sometimes, a range of mountains. The conditions for a 'superbloom' require a devilishly intricate set of circumstances. The soil must be dry enough to avoid invasive grasses. The autumn rainfall must be deep enough to penetrate into the soil and reach the dormant wildflower seeds but not so much as to result in run-off or flash floods that carry the seeds away. Then, over winter and early spring, there must be cloud cover to shield the soil from intense heat or overnight cold, and, as the plants start to grow, the wind must not be too strong. Perhaps, given all that, it is predictable that superblooms occur unpredictably and relatively rarely. Desert superblooms in California are estimated to occur on average every ten years or so but, of course, no one really knows until the hills come alive.

The power of flower

Natural superblooms have become, in the age of Instagram, a magnet for tourism, which comes with its own set of issues, not least that the fragile desert landscape, even when dressed in its Sunday best, cannot cope with a human stampede. In March 2019, an estimated 50,000 people descended on Walker Canyon near Lake Elsinore in California to see and be seen among the orange sea of California poppies (*Eschscholzia californica*). What had been hailed as a magically thrilling event on social media as a #poppy-paluza became, as the crowds came and the traffic jammed the roads, a #poppy-nightmare. A similar surge in wildflower mania in 2017 in the town of Borrego Springs, also in California, was dubbed 'Flowergedden' as supplies of petrol, food and bathrooms ran out. It is a fact of superbloom modern life that often the people who come to see and surround themselves with the flowers end up trampling on more than a few.

Eschscholzia californica

Where to see a natural superbloom

1. **Namaqualand, South Africa**: Between late August and October; daisies, vygies and salvias.

2. **Death Valley, USA**: Early spring, most usually March; sunflowers, desert gold (*Geraea canescens*), gravel ghost (*Atrichoseris platyphylla*).

3. **Anza Borrego Desert, California, USA**: Early spring, March; desert sunflowers (*Geraea canescens*) and lilies (*Hesperocallis undulata*), brown-eyed primrose (*Chylismia claviformis*).

4. **Atacama Desert, Chile (including Desierto Florido National Park)**: Mid-September to mid-November; *Cistanthe grandiflora* and *Bomarea ovallei*.

5. **Antelope Valley, California, USA**: Early spring, usually March; orange California poppies (*Eschscholzia californica*).

6. **Western Australia**: Starts in the north in June and moves south as it turns to spring in November; everlasting daisies, tall mulla mulla (*Ptilotus exaltatus*), kangaroo paws (*Anigozanthos* spp.), banksia.

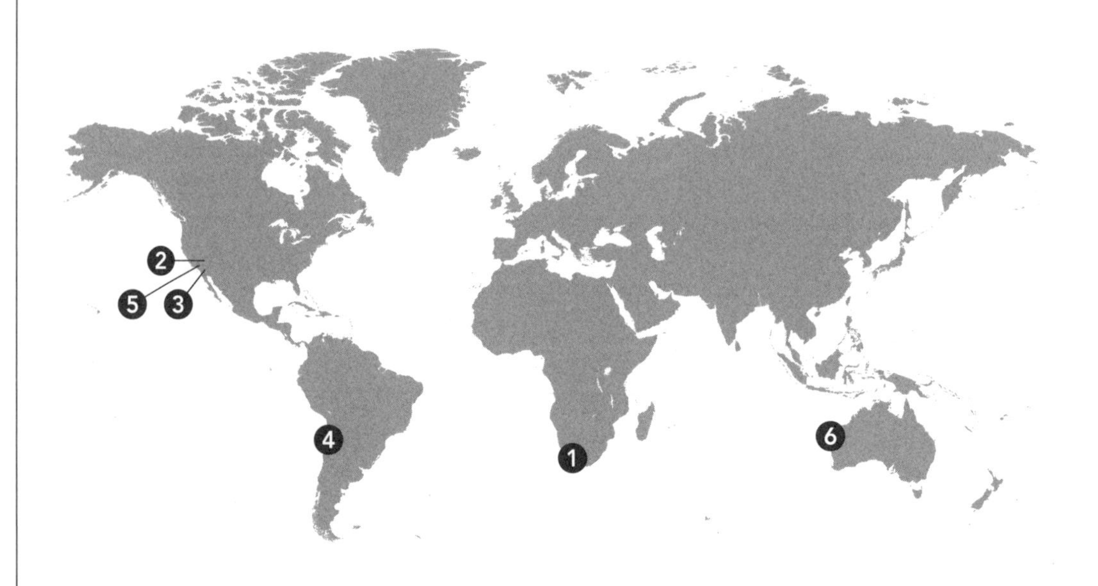

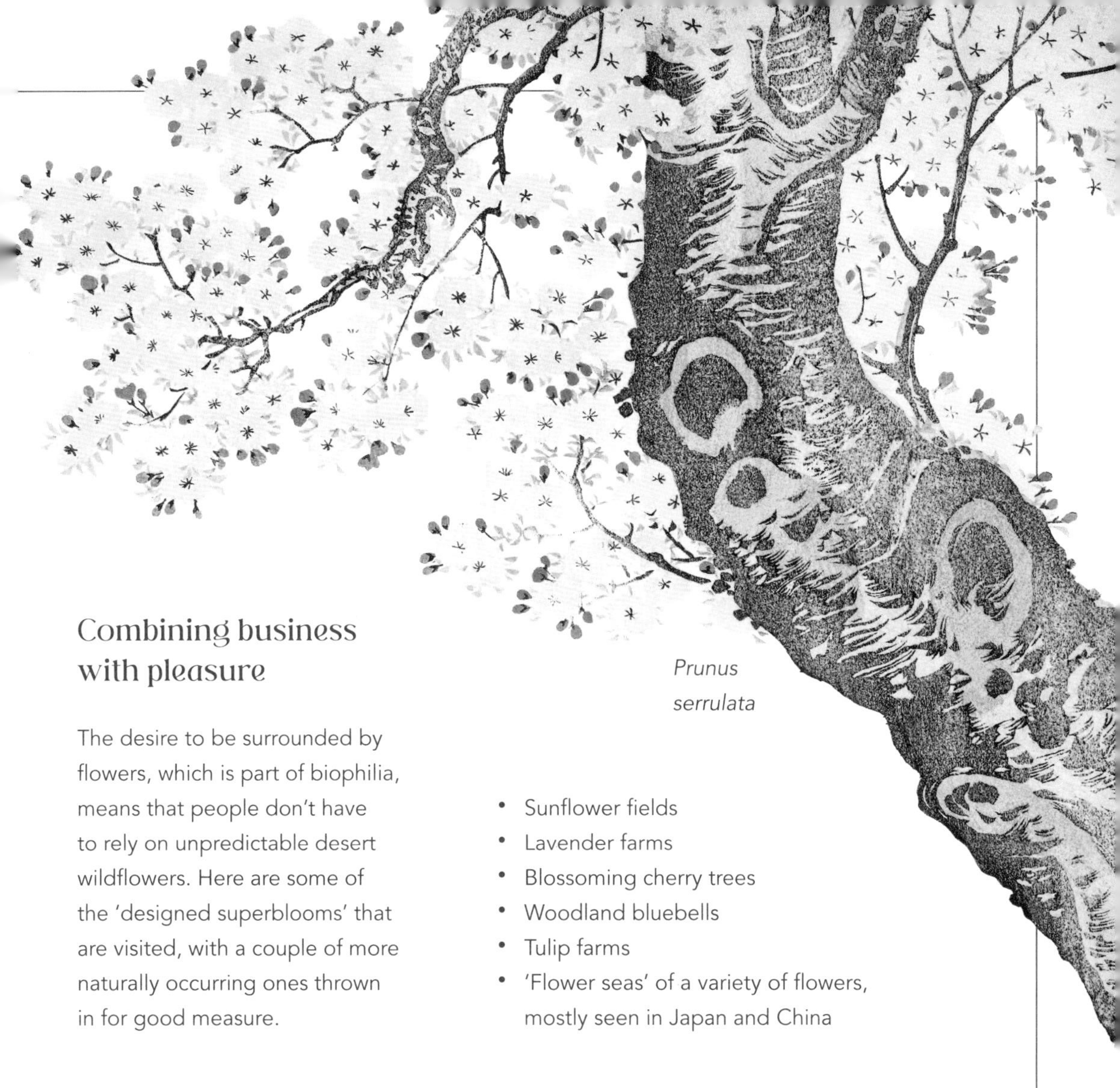

Prunus serrulata

Combining business with pleasure

The desire to be surrounded by flowers, which is part of biophilia, means that people don't have to rely on unpredictable desert wildflowers. Here are some of the 'designed superblooms' that are visited, with a couple of more naturally occurring ones thrown in for good measure.

- Sunflower fields
- Lavender farms
- Blossoming cherry trees
- Woodland bluebells
- Tulip farms
- 'Flower seas' of a variety of flowers, mostly seen in Japan and China

'Sauntering in any direction, hundreds of these happy sun-plants brushed against my feet at every step, and closed over them as if I were wading in liquid gold.'

Scottish-American naturalist John Muir, *The Mountains of California*, 1882

It's a small world

Small, small world: bonsai

Bonsai, which translates as 'planted in a pot', has its roots in ancient Chinese horticulture, but it was in Japan where it flowered (sometimes for real) into something truly special under the influence of Zen Buddhism. At its core is the idea that, through pruning and care, you can create a miniature version of a tree and, in the process, enhance your own life. Pine, maple, juniper and cherry are popular choices for bonsai, which have a variety of ideal shapes or styles. Creating a bonsai teaches patience, self-awareness, consistency, compassion and creativity. Quite a large package for a very small tree.

Formal upright
Chokkan

Laterati or Bunjin
Bunjingi

Slanting
Shakkan

Windswept
Fukinagashi

Twin trunk
Sokan

Informal upright
Moyogi

Twisted
Nejikan

Exposed roots
Neagari

Bonsai-in-rock
Ishisuki

Root-over-rock
Seki-joju

Cascade
Kengai

Semi-cascade
Han-kengai

Broom
Hokidachi

Multiple trunk
Kabudachi

Forest
Yose-ue

Even smaller world: terrarium

Terraria are small spaces that allow you to garden in another world where plants, and landscapes, come in miniature. They can either be left open to the air, for a display of cacti and succulents, or closed with a lid (or cork), which allows you to create a small tropical world. A terrarium needs to be situated where there is plenty of natural light but no direct sunlight. Once made, spray it (inside) with water every few weeks or when the soil is dry. If it becomes too wet (condensation on the glass), open the lid (if it has one) for a few hours to clear it.

You'll need:

- A glass jar, any size, with or without lid, according to plant choice
- Small stones such as gravel or small pebbles
- Potting soil (organic and peat free), a few handfuls
- Activated charcoal (keeps water fresh and stops bacteria)
- Small tools such as a pencil or chopstick, tiny trowel or long spoon
- Optional: Sphagnum moss, faerie equipment, glass animals, other miniature decorations

Plants for an open terrarium

- Air plant (*Tillandsia stricta*)
- Crassula
- Mexican snowball (*Echeveria elegans*)
- Haworthia

Tropical plants for a closed terrarium

- Nerve plant (*Fittonia albivenis*)
- Polka dot plant (*Hypoestes phyllostachya*)
- Peperomia, including 'Pepperspot'
- Earth star (*Cryptanthus bivittatus*)

1. Wash and dry your container.
2. Add a 2–3cm (¾–1in) layer of gravel or pebbles.
3. Scatter charcoal over the top.
4. Add sphagnum moss (optional).
5. Add potting soil, at least as deep as the largest rootball of your plants.
6. Make holes in the soil with a pencil or tool.
7. Add your plants, largest first. Do not place too close to the glass.
8. Tamp down the soil around the plant with a long spoon or end of a pencil.
9. Add any top decoration (moss, small coloured stones etc).
10. Spray the inside of the terrarium with water. Place the lid (or not).

And, finally, the smallest: kokedama

Kokedama translates as 'ball of moss', which underplays it slightly, as this is a ball of soil with one plant, with the whole thing covered in moss. The idea comes from Japan but is now trendy throughout much of the houseplant-loving world. You can pop your kokedamas on a tray or plate, but by far the most impressive way to display them is as a string garden, with each of them hanging down, separate but together, from the ceiling. Success depends on placement (not in direct sun or in a draught) and your choice of plant. The enemy is dry air. It's obvious, but you don't want a plant that is going to grow too big. Small succulents can work well, but perhaps the best choice would be epiphytic plants, such as the smaller bromeliads or rabbit's foot ferns, as they hang around in mid-air naturally.

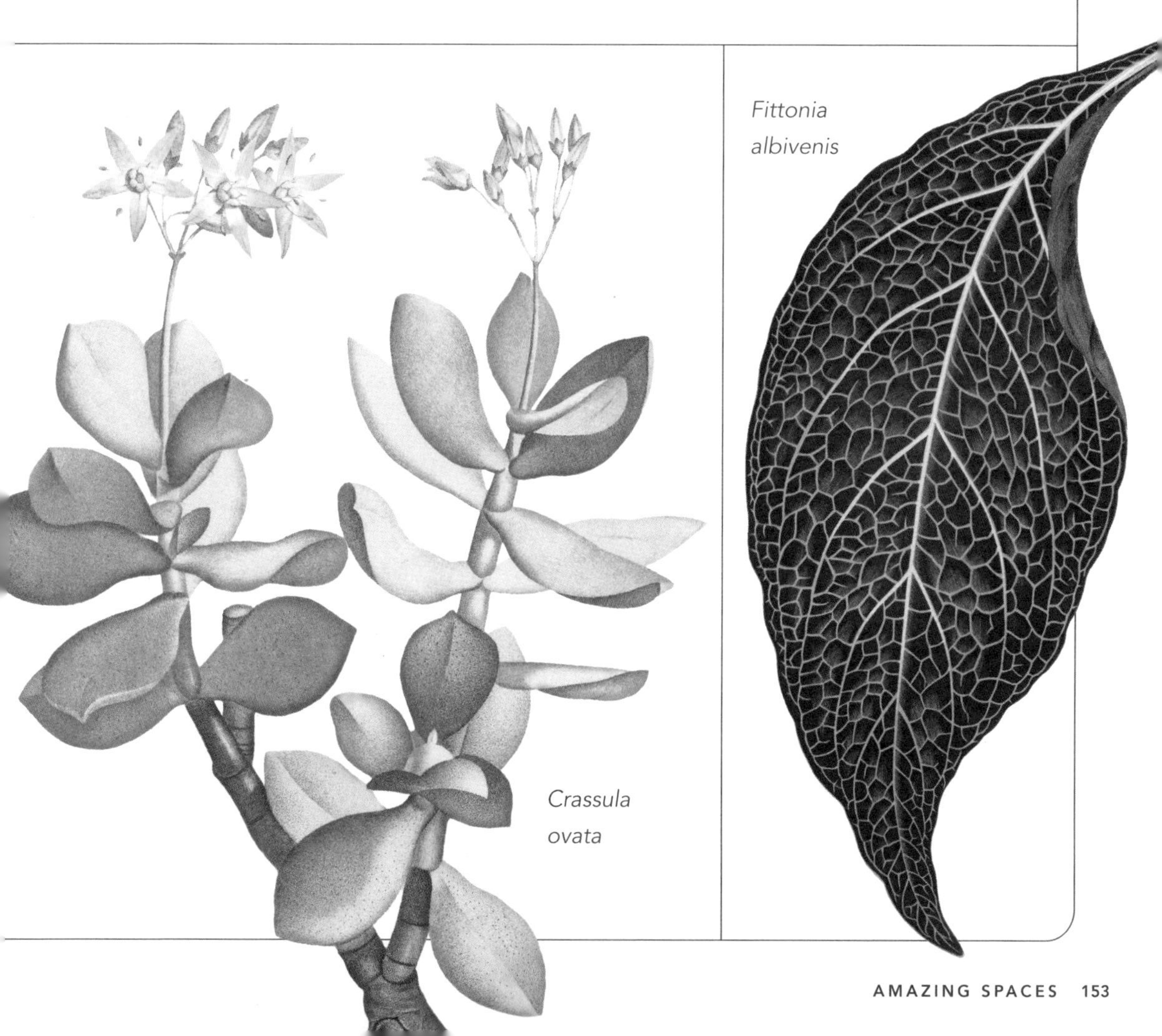

Crassula ovata

Fittonia albivenis

Gardens in the sky

A walk on the Wild West Side

The last train that ran on what was the West Side Elevated Line in Manhattan, New York City, USA, was in 1980, then, for decades, the tracks just sat there, an eyesore, abandoned. Demolition loomed. Then something remarkable happened: where almost everyone saw debris and chaos, others saw beauty. 'When I first stepped up on the High Line in 1999, I truly fell in love,' writes Robert Hammond, who lived nearby. 'What I fell in love with was the tension. It was there in the juxtaposition between the hard and the soft, the wild grasses and billboards, the industrial relics and natural landscapes, the views of both wildflowers and the Empire State Building. It was ugly and beautiful at the same time. And it's that tension that gives the High Line its power.'

Hammond and his friend Joshua David founded the Friends of the High Line and the original idea was to do nothing – literally. Leave it to the wildflowers and grasses. But practicalities meant the structure had to be remediated and new drainage put in. All the plants – and rails – had to be taken up. Plan B had to be devised. It was the planting design of Dutchman Piet Oudolf that inspired them (and still does).

His plan, with a multi-season garden of perennials, kept the original tension. Eric Rodriguez, the director of horticulture for the park, describes the planting as being like a vertical sandwich. The bottom layer is a grass matrix, topped by herbaceous perennials, with some areas also having shrub and tree canopy layers. Trees must be chosen carefully, as the

soil is only 46cm (18in) deep, which also explains why so much of the 2.5-km (1½-mile) linear park that is 9m (30ft) above the ground is given over to herbaceous plants.

The planting is dense, with six to eight plants per 30sq cm (1sq ft). This is partly what gives it its wild nature and, more practically, reduces the need to water and smothers weeds. There are some 400 plant species and each has to be hardy enough to survive in this harsh environment, open to wind and weather, not to mention viewed by millions of visitors a year. The park is owned by the city but run and maintained by the Friends of the High Line and volunteers. Over the seasons, the top growth is allowed to die and go brown. All the 100,000-plus plants are pruned and trimmed during the spring cutback, just in time for the bulbs to start peeking through.

The High Line has been the inspiration for many cities to reassess their rusting disused infrastructure.

More sky parks and gardens

- Bloomingdale Trail, Chicago, USA
- Seoullo 7017, Seoul, South Korea
- Coulée Verte René-Dumont, Paris, France
- Castlefield Viaduct, Manchester, England, UK
- Sky Garden, London, England, UK
- Warrumbungle Dark Sky Park, New South Wales, Australia
- Randalstown Viaduct Community Garden, Northern Ireland, UK

Garden follies

Definition

Folly (*noun*): a building in the form of a castle, temple, etc, built to satisfy a fancy or conceit, often of an eccentric kind.
Collins Dictionary

Surely this is pure folly?

It was Roman statesman Marcus Tullius Cicero who said, 'If you have a garden and a library, you have everything you need.' But Chanticleer Garden, near Philadelphia, Pennsylvania, USA, has gone one step further: it has placed a library *in* a garden, complete with books and seating. Of course, it's a folly, and a totally remarkable one. The library is one of several 'rooms' in the Ruin Garden which was constructed in the early 2000s on the footprint of an old house. Climbers and espaliered trees green the partial walls and open windows. There is a 'great hall' with a mantelpiece and a large table that is a reflecting pool, and a library complete with stone books labelled as 'Moss' and 'Ex Libris'. The planting is almost completely silver and green. It feels very real – and unreal at the same time. For a folly, that is a major success.

Famed follies

- **Orcus monsters**, Garden of Bomarzo, Lazio, Italy, 1552
- **Conolly's Folly**, County Kildare, Ireland, 1740
- **Hameau de Chantilly**, Château de Chantilly, Oise, France, 1774
- **Roman ruins**, Schönbrunn Palace, Vienna, 1778
- **Parc Monceau**, Paris, France, 1779
- **Roman aqueduct**, Arkadia Park, near Nieborow, Poland, 1784
- **Broadway Tower**, near village of Broadway, Worcestershire, England, UK, 1798
- **Belvedere Castle**, Central Park, New York City, USA, 1869
- **Parthenon**, Nashville, Tennessee, USA, 1897

Build a faux ruin

A fake ruin is not allowed to have an obvious functional purpose, but that doesn't mean it can't have a horticultural one. It can be a focal point, create a shady microclimate or help to divide a garden into 'rooms'. It can add height, intrigue and depth to a garden. If you are planning on using a bit of it as a storage area, for instance, then you must do this only secretly and tell no one. You can buy fake ruin kits but it's much more fun to make your own.

1. Find or acquire building materials. It helps to have a load of stones (dug up from your garden perhaps) or unwanted bricks from a nearby building site. Rubble is ideal for making a foundation.
2. Design with fakery in mind. Don't just build a wall; instead build a wall with an empty window shape in it (literally to frame a view) and, to add depth, some kind of corner.
3. Build it well but not too well. It can't fall down but it can't look perfect either.
4. Do something to make it look older instantly. Buy yoghurt in bulk and paint it on: the bacteria attracts all sorts and ages it.
5. Moss, lichen and ivy instantly add years.
6. Stand back and admire.
7. Let weeds colonize it naturally.

Part of an aqueduct in Rome

Living architecture

A vertical forest – with flats attached

It may have seemed a crazy idea to plant a vertical forest in 'green balconies' surrounding high-rise apartment blocks, but that didn't deter Italian architect Stefano Boeri. 'We have to multiply the number of trees everywhere. And the reasons are very clear. It's a faster, cheaper and more inclusive way to try to take down global warming,' he said. Thus, in 2014, Bosco Verticale was born: two towers (112m/367ft and 80m/262ft, 26 and 18 storeys) in Milan's Porta Nuova district. They house 113 apartments, 800 trees, 5,000 shrubs, thousands of smaller plants and thousands of birds and butterflies.

'Green design' makes nature a core component of a building, not an incidental add on. Bosco Verticale's trees and plants, watered by an integrated (refiltered) irrigation system, are said to convert an average of almost 20,000kg (44,000lb) of carbon each year. They also absorb pollution, minimize street noise, lessen heat-island impact and attract nature, as shown by the birds that nest there now.

The trees were the first occupants of the building, with the humans following, and the buildings have become wildly popular. The trees are mostly deciduous, providing a constantly changing colour palette. The forest and plants are tended by 'vertical gardeners' – think high-rise window cleaners but with secateurs – who are both arboriculturists and expert climbers. They garden by abseiling from the roof and are known as the 'flying gardeners', adding new dimensions to horticultural skill-sets.

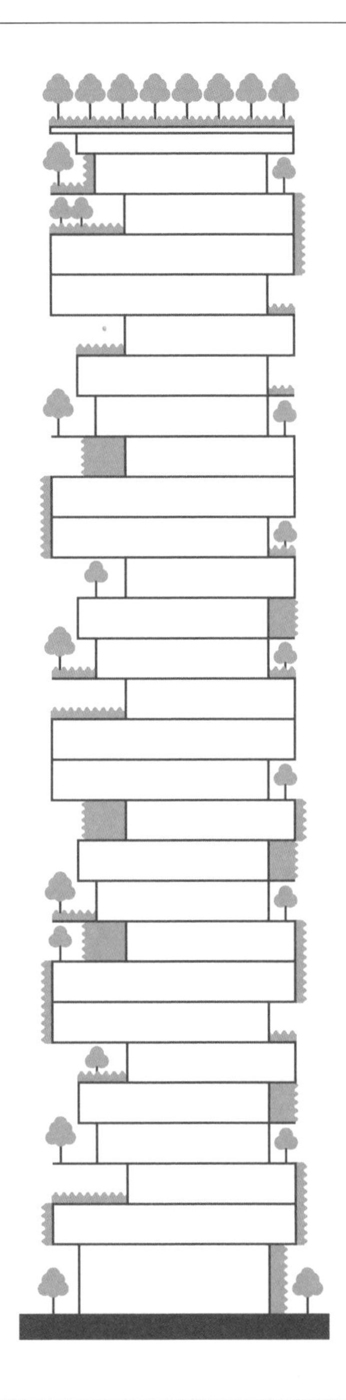

Other living buildings

One Central Park, Sydney, Australia

Features two residential towers plus mixed-use retail area, opened in 2014. 'Green' aspects include vertical hanging gardens, a cantilevered heliostat to provide reflected light to plants, and an internal water-recycling plant. The vertical gardens, which use a hydroponic system designed by botanist Patrick Blanc, thus avoiding the weight of soil, stretch up to 50m (164ft) high. There are 23 green walls, and both exotic and Australian native plants make the living walls.

Valley, Amsterdam, The Netherlands

The three towers of 67m (220ft), 81m (266ft) and 100m (328ft) have a geologically inspired design, combined as they are to look like a craggy canyon. Multi-use, with 370 vegetated areas and retail, eating, offices and cultural areas, plus 200 apartments. As it matures, the vertical forest will emerge.

The Urban Farm, Level 51, CapitaSpring, Singapore

The entire city of Singapore is committed to the idea of covering skyscrapers and high-rise buildings with plants. They call it Skyrise Greenery and the goal is to become a City of Nature, using plants to provide climate, ecological and social resilience in a mega-urban environment. The 3,000sq m (10,000sq ft) 'urban farm' or 'food forest' on the top floor of the CapitaSpring skyscraper provides a taste, in every way, of this vision. It has five themed gardens – Singapore Food Heritage, Wellness, Mediterranean Potager, Australian Native and Japanese Potager.

Lantana camara, a naturalized shrub species found in CapitaSpring, is a host plant for the Pygmy grass blue, the smallest butterfly in Singapore.

Back to the 17th-century home garden

The useful garden

The 17th-century home garden was a combination greengrocer, chemist, florist and general store with a hint of haberdashery. The garden at even a labourer's cottage would contain plants to provide food and seasonings for three meals a day, cures for everything from headaches to piles, material for clothes, and deterrents for pests. Plants such as fennel would be known by everyone – as *Culpeper's Complete Herbal* (1653) notes: 'Most gardens grow it; and it needs no description.'

Foeniculum vulgare

Veg of the era

These would include beans, carrots, cabbages, gourds, parsnips and pompions (pumpkins). A staple was the pea, which would not have been the unripe green pea but the Carlin pea, also called black peas, grey peas, maple peas and black badgers. Traditionally, these were eaten as parched peas, which meant boiled with lashings of malt vinegar, plus salt and pepper. Carlin peas were also dried and ground to make a kind of flour, and the basic ingredient for pease porridge, which, on special occasions, may have had rashers of bacon added, though often only herbs that somehow acted as faux meat.

Pease porridge hot
Pease porridge cold
Pease porridge
in the pot
Nine days old

Traditional nursery rhyme

Helpful herbs

Herbs for cooking: A wide variety of culinary herbs included staples such as parsley, sage, rosemary and thyme. Others included those used to season pease porridge, such as Good-king-Henry (*Chenopodium bonus-henricus*) and fat hen (*Chenopodium album*), which was also known as 'bacon weed' and 'lamb quarter'.

Herbs for drinks: Water was not seen as safe. Homemade drinks included ale, cider and mead. To hide the sour taste of ale, aromatic herbs known as 'gruiting' would have been added, which included bog myrtle (*Myrica gale*), mugwort, horehound (*Marrubium vulgare*), heather (*Calluna*), ground ivy (*Glechoma hederacea*, aka alehoof) and alecost or costmary (*Tanacetum balsamita*).

Medicinal herbs: A herbal 'Simple' was a home remedy that involved only one ingredient, though that ingredient could be used as a simple for more than one ailment. Thus, you would use feverfew (*Tanacetum parthenium*) for a headache but also for arthritis, tinnitus and other problems including as a 'general strengthener of the womb' (Culpeper). Medicinal herbs included fennel, comfrey, plantain, marjoram, sweet cicely (*Myrrhis odorata*), tobacco and many others. Compound cures used more than one plant (apothecaries sometimes used a dizzying array of herbs).

Salvia officinalis

Thymus vulgaris

Snail water recipe

Herbs could be combined with other, earthier items, including snails and worms (roasted and ground up in some cases). Margaret Willes, in her book *The Domestic Herbal* (2020), shares this 1660 recipe from the Duchess of Lauderdale. The treatment was to be drunk three times a day (morning, afternoon, night) and was particularly good for 'weak children and old people'.

'Take a pottle of Snails and wash them well in 2 or 3 waters, and then in Small Beer, bruise them, shells and all, then put them into a gallon of Red-cows milk, Red rose leaves dryed, the whites cut off Rosemary, Sweet Marjoram, or each one handful, and distil them in a cold still, and let it drop upon powder or white sugar candy.'

Clothing-related plants

Hemp (*Cannabis sativa*) and flax (*Linum usitatissimum*): Cloth

Soapwort (*Saponaria officinalis*): Soap

Dyeing plants included: woad (*Isatis tinctoria*), madder (*Rubia tinctorum*), dyer's greenweed (*Genista tinctoria*), galls (aka oak apples)

Cuckoo pint (*Arum maculatum*), also known as starch plant: Starch for collars (lots of ruffs in Elizabethan times)

Arum maculatum

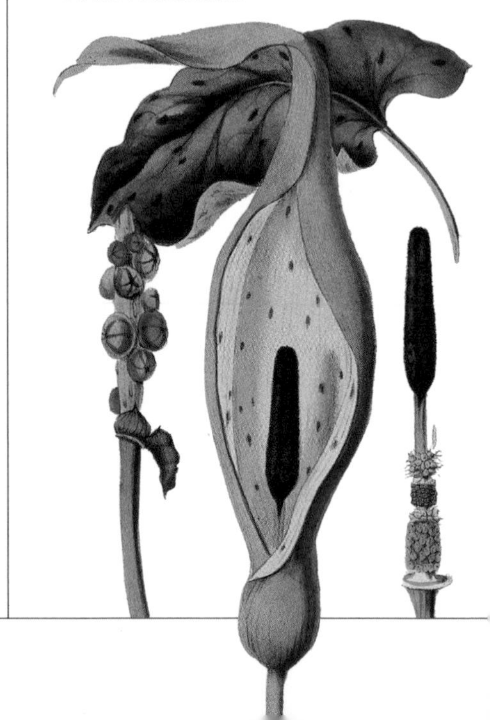

Herbal household cleaners

Strewing herbs, to be placed on the floor, to be walked on, releasing scents, were sedges and rushes including meadowsweet (*Filipendula ulmaria*) and sweet flag (*Acorus calamus*).

To control fleas: fleabane (*Pulicaria dysenterica*). Strewn alder (*Alnus glutinosa*) leaves were believed to attract fleas, and then, after the fleas were suitably ensconced, the whole lot was swept out.

To deter moths: rue (*Ruta graveolens*), wormwood (*Artemisia absinthium*), southernwood (*Artemisia abrotanum*), bog myrtle (*Myrica gale*, aka sweet gale).

To ward off mice: mint (*Mentha*).

Ruta graveolens

'Have millions [melons] at Mihelmas, parsneps in lent: In June, buttred beanes, saveth fish to be spent. With those, and good pottage enough having than: Thou winnest the heart, of thy laboring man.'

English poet and farmer Thomas Tusser, *A Hundreth Good Pointes of Husbandrie* (1557)

Myrica gale

A to Z of notable gardens

Anne Spencer House and Garden Museum, Lynchburg, Virginia, USA: Home and garden of Harlem Renaissance poet Anne Spencer (1882–1975). The only known restored garden of an African American woman. The garden includes her writing cottage, Edankraal, named after her husband Edward and her name, as well as the African word *kraal* for 'enclosure' and/or 'community'.

Butchart Gardens, Brentwood Bay, British Columbia, Canada: Former limestone quarry transformed into a magnificent 22-hectare (55-acre) sunken display garden.

Claude Monet's garden, Giverny, France: Impressionist master's garden, with fabulous flower-filled paths and the famous water lily pond (see page 51).

Dubai Miracle Garden, Dubai, UAE: Largest flower garden in the world (72,000sq m/24,000sq ft) with 50 million flowers and 250 million plants.

Eram Garden, Shiraz, Iran: Elegant historic garden on the northern shore of the Khoshk River that originally dates from the 12th century.

Folly Farm, Sulhamstead, Berkshire, England, UK: Iconic Arts & Crafts home and garden transformed between 1906 and 1912 by architect Sir Edwin Lutyens and garden designer Gertrude Jekyll. A new redesign of the garden by Dan Pearson has been saluted as referencing Jekyll while also introducing new areas, including a wind garden of grasses.

Mormodes luxata, an orchid native to Mexico and likely to be seen at Jardin Botanico de Vallarta.

Gardens by the Bay, 'Supertree Grove', Singapore: Futuristic structures that dominate the skyline and contain vertical gardens showcasing a gorgeous mixture of tropical plants.

Humble Administrator's Garden, Suzhou, China: One of the top classical gardens in China and listed as a World Heritage Site by UNESCO.

Ilm, or Park an der Ilm, Weimar, Germany: Large landscape park influenced by Johann Wolfgang von Goethe – his 'garden house' is inside the park.

Jardin Botanico de Vallarta, Puerto Vallarta, Mexico: Tropical highland garden that is home to Mexico's most extensive orchid collection.

Keukenhof, Lisse, The Netherlands: Vibrant flower garden covering 32 hectares (79 acres) with an amazing 7 million bulbs that are planted annually.

Longwood Gardens, Pennsylvania, USA: One of the world's largest display gardens is a series of woodlands, meadows and gardens, including an outstanding topiary garden dating from the 1930s.

Majorelle Garden, Marrakesh, Morocco: Walled garden with dramatic planting and zingy colours including acid yellow, hot red and the vivid blue that is now known as Bleu Majorelle. Created from the 1920s by painter Jacques Majorelle, it was bought and restored in 1980 by French fashion designers Yves Saint-Laurent and Pierre Bergé.

Nong Nooch Tropical Botanical Garden, Chonburi, Thailand: A 200-hectare (500-acre) garden and tourist attraction that has a scientific centre devoted to the study of cycads.

Organopónico Vivero Alamar, Havana, Cuba: Cuba's most successful urban farm covering 10 hectares (26 acres) in the Alamar district. Provides organic fruit and veg, as well as employment.

Powerscourt Estate, Enniskerry, County Wicklow, Ireland: Showpiece garden featuring striking views, terraces and a memorable woodland garden.

Quinta de Monserrate, Sintra, Portugal: Romanticism-inspired landscape in the fairytale-worthy surroundings.

Ryoan-ji Temple, Kyoto, Japan: Zen *karesansui*, or dry landscape rock garden, dating from the late 15th century. The relatively small garden, 25 x 10m (82 x 33ft) has 15 stones of different sizes, in groups, placed in white gravel raked daily by monks. The only plant life is moss.

Shalimar Bagh, Srinagar, Jammu and Kashmir, India: Stately and extensive 17th-century *charbagh* that is the most celebrated of the royal gardens in India.

Tasmanian Botanical Gardens, Hobart, Australia: Australia's cool-climate garden that contains the only sub-Antarctic plant house in the world.

Encephalartos horridus

University of Oxford Botanic Garden, England, UK: Founded in 1621, this is one of the oldest scientific gardens in the world. It has more than 5,000 plant species, including a notable modern medicinal plant collection.

Villa d'Este, Tivoli, near Rome, Italy: This innovative garden is a showcase of Renaissance culture and was the first of Italy's *Giardini delle Meraviglie* (Gardens of Wonder). It is set out over two steep slopes, with an astonishing array of gravity-led water cascades, fountains, waterfalls, jets, pools and grottoes. A masterpiece of garden design – and hydraulic engineering.

Wigandia, Noorat, Victoria, Australia: Modern dry garden created by William Martin on the slopes of an extinct volcano, with an outstanding and eclectic collection of plants.

Xilitla, Las Pozas, San Luis Potosi, Mexico: Surrealist sculpture garden created by Edward James in the subtropical rainforest of the Sierra Gorda mountains of Mexico.

Yuyuan Garden, Shanghai, China: Classical walled garden, with many period buildings, that has survived amid the neon and noise of Shanghai.

Zahrady Pražského Hradu, Prague, Czech Republic: Renaissance castle gardens located in the centre of the city, on the site of medieval vineyards.

Rhododendron canescens

Chapter 5

Inspiring People

Creative. Innovative. Brave. This chapter celebrates people who changed – and are changing – the world of plants and gardens by believing in something and then doing it with fortitude and passion. It's our own hall of fame, which includes tales of visionaries, explorers, designers and dreamers.

Women who broke the mould

'Artist, Gardener, Craftswoman'

Gertrude Jekyll is often cited as one of the most influential gardeners of all time, but it is interesting to note that, if she had had her way, she wouldn't have been a gardener at all. Jekyll, born in 1843, wanted to be a painter in an age when women did not, in general, aspire to such things. Jekyll was singular: well educated, widely travelled, resolutely single and without children. She was one of the first women in Britain to train as a painter and was also an accomplished textile artist. She only turned to gardening when her eyesight became too poor, because of her extreme myopia, for painting or embroidery. Thus she was almost 40 when she designed her first garden. In 1889 she met a young architect named Edwin Lutyens over a cup of tea in Surrey (they talked about rhododendrons, apparently). She hired him to design her home at Munstead Wood while she did the garden. They would become a great partnership, a signature duo of the Arts and Crafts movement, but they made an unlikely pair: she the redoubtable middle-aged spinster; he the young and untried architect who called her 'Aunt Bumps'. They would work together on some hundred homes and gardens – creating magical, romantic

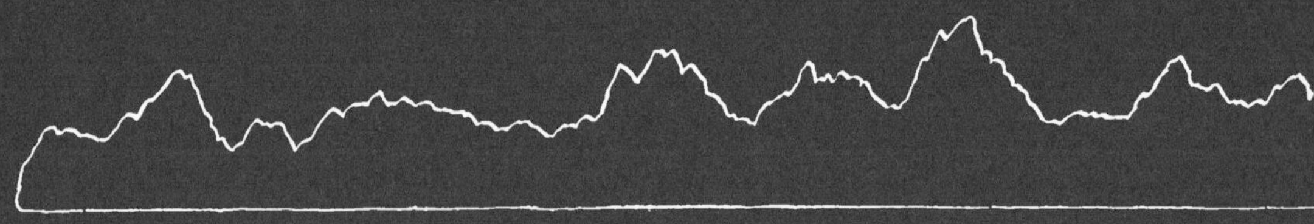

ELEVATION HEIGHT-LINE OF BACK PLANTS

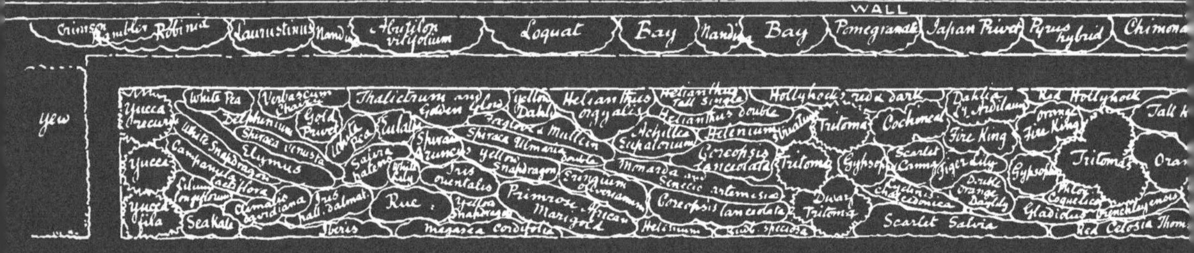

spaces that still enchant today. Jekyll, wildly productive, would design at least 300 more gardens, most of the time working from home.

If it all sounds rather overly English, her gardens have none of the Victorian or Edwardian fussiness that implies. Jekyll had the eye of a painter, and she designed borders in swathes and waves of colour that flowed from cool to hot and (sometimes) back again. She was also practical: she had a nursery and sold plants to her clients, but only after she tried them out herself. Jekyll was a gifted writer, even though she did not publish her first book until the age of 56. Her writing on gardening still seems modern and her turn of phrase can be brilliant. In her 1908 book *Colour in the Flower Garden*, she describes a particular scarlet for a geranium: 'The colour is pure and brilliant but not *cruel*.' She writes about how we should aspire to create 'pictures' and 'scenes' in our gardens. She died in 1932, and her gravestone, designed by Lutyens, says simply: 'Artist, Gardener, Craftswoman'.

'I am strongly of the opinion that the possession of a quantity of plants, however good the plants may be themselves and however ample their number, does not make a garden; it only makes a collection. *Having got the plants, the great thing is to use them with careful selection and definite intention. Merely having them, or having them planted unassorted in garden spaces, is only like having a box of paints from the best colourman or, to go one step further, it is like having portions of these paints set out upon a palette. This does not constitute a picture; and it seems to me that the duty we owe to our gardens and to our own bettering in our gardens is so to use the plants that they shall form beautiful pictures.'*
Gertrude Jekyll, *Colour in the Flower Garden* (1908)

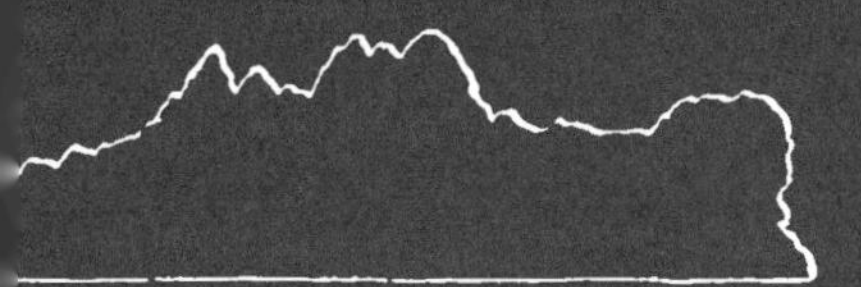

Plan of the main flower border, illustration from Gertrude Jekyll, *Colour in the Flower Garden*, 1908

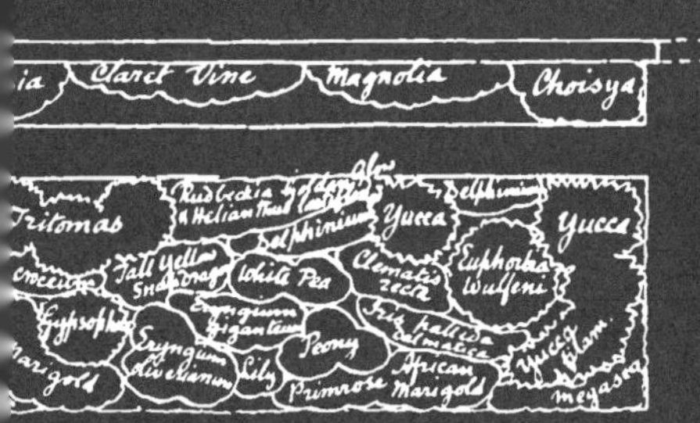

Other ground-breaking female gardeners

Beatrix Farrand
(1872–1959):
New Yorker with an artistic eye who became a trailblazer for women in landscape architecture and garden design.

Vita Sackville-West
(1892–1962):
Poet, writer and inspirational amateur gardener who created the famed White Garden at her home in Sissinghurst, Kent, England, UK.

Ruth Bancroft
(1908–2017):
Expert on succulents who set up a permanent xeriscape garden near San Francisco, California, USA.

Beth Chatto
(1923–2018):
Superb and innovative plantswoman who created a world-famous gravel garden in Essex, England, UK.

Breadfruit teaches history

The Antiguan-American novelist and academic Jamaica Kincaid, born in 1949, loves her overflowing garden in Vermont, in New England, USA, for a multitude of reasons. She has written, extensively, about plants and her garden, which she has tended for some 30 years. 'The garden has taught me to live, to appreciate the times when things are fallow and when they're not.'

When Kincaid looks at a garden, she sees something beyond the border or, even, the horizon. For her, plants can also be freighted with a troubled history, and she is not afraid to confront the more difficult aspects of the subject. She grew up in Antigua, where, she notes, many of the plants were not native at all but brought there by the British. Breadfruit (*Artocarpus altilis*) is just one example, sent to the West Indies to feed slaves.

Cotton was one of the main crops of Antigua and, in America, it is a plant closely associated with the history of slavery. But in her garden in Vermont she grows it because she loves the flower. 'Every year I have some grief with it and I will sometimes say: "It's clear that I'm not meant to grow cotton as an amusement. It's forbidden that a Black person should grow cotton just because she loves the flowers."' The dark history of plants is something that is explored in *An Encyclopedia of Gardening for Colored Children* (2024), produced with artist Kara Walker.

'The tamarind, the mango, hibiscus . . . all these things that were part of my childhood imagination were interwoven with empires, and the empire is violent and political.'
Jamaica Kincaid

Artocarpus altilis

Victorian plant hunters

Swashbuckling escapades

Plant hunters during and after the reign of Queen Victoria are often talked about as real-life versions of Indiana Jones, and their escapades, as they combed the world for 'new' species, are indeed incredible. Danger, disease, disaster – it was all part of the job in the age of empire, a world of cutthroat competition, ruthless exploitation and daring exploration. It was also a very male world.

The woman who slept on a volcano

The *Werneria staffordiae* is a Peruvian sub-shrub that is named after the only female plant hunter of her time. Her name was Dora B. Stafford and there is little trace of her left today. There is no biography, long nor short, and it is only via an entry in the Harvard list of botanists that we know that Stafford, from north London, was collecting plants in deepest Peru from 1930 to 1939. She did give an interview to the *Daily Herald* in 1938 that was entitled 'She sleeps on a volcano!' and in which she is introduced as: 'Fair-haired Miss Dora Stafford, only woman plant-hunter, is back in London after 15 months' adventuring in southern Peru. She tells of the thrills and perils on the remote awe-inspiring heights of the Andes where she risks her life seeking – rock plants.' Stafford told the paper about how she had avoided wild dogs, bandits and volcanic eruption as she collected her alpine specimens (she had some 1,000 specimens, including many gentians, a great number of which went to Kew). In 1938, in an interview with the *Ellesmere Guardian* in New Zealand, she gave this rather unusual insight about what it's like to live and work high up in the Andes: 'One advantage of the great height was that food keeps fresh and good for an extraordinary long time. An egg, for instance, is still perfectly good a month after being laid.' Dora Stafford has been forgotten in Europe but in Peru her name lives on in the sub-shrub that thrives, appropriately, on mountain slopes.

The long, tall tale of the handkerchief tree

The next time you see a handkerchief tree (*Davidia involucrata*) in full blowsy bloom, spare a thought for Ernest Henry Wilson. He was only 23 years old when, in 1899, he accepted a job as a plant hunter with Veitch & Sons in England, UK, with these specific instructions: 'The object of the journey is to collect a quantity of seeds of a plant the name of which is known to us. This is the object. Do not dissipate time, energy or money on anything else.'

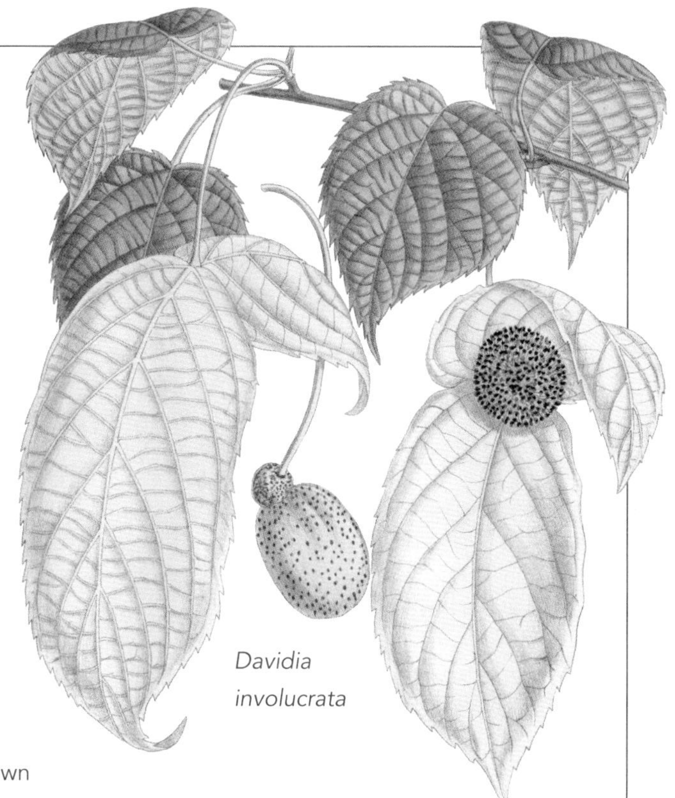

Davidia involucrata

The plant was *Davidia involucrata*, known then as the dove tree. It was an epic journey, by boat and train, to America and then onward to southwest China, dodging bubonic plague and various outbreaks of hostilities. Wilson, who later wrote about this in his 1907 book *Aristocrats of the Garden*, finally managed to visit the man (in China) who had, a few years earlier, seen this one tree. He sketched out the location, one point on a map of a geographical area the size of the state of New York.

Wilson's journey in the remote Yunnan region included a brush with bandits, imprisonment for spying and near drowning on a turbulent river. Incredibly, he did manage to find the tree, only to find it was now, well, a stump, with a newly built house nearby. But a few months later, gathering seeds elsewhere, he found the same tree again, this time in full flower ('with its wealth of blossoms more beautiful than words can portray.') But it was May; he had to wait until November to gather the seed.

Wilson did bring the seeds back to England, though he notes, through the written equivalent of gritted teeth, that, as it turned out, a Frenchman had already had the seeds and grown the tree, so took the credit. Wilson brought back some thousand seeds of 'new' plants from the Far East in his time (his nickname was 'Chinese'), but the handkerchief tree was not his glory.

Ribes sanguineum

Mr Douglas, of fir fame

David Douglas was born in Scone, near Perth, Scotland, UK, in 1799, a natural rebel who loved adventures, eagles, the outdoors and plants (possibly in that order). By the age of 11 he was working as an apprentice gardener at Scone Palace and, after seven years, continued his education and impressed his bosses enough to be recommended to the Horticultural Society of London (now the Royal Horticultural Society) as an ideal plant hunter. Douglas was a larger than life adventurer, with his pet eagles and tartan outfits worn as he gathered seeds in the wilderness forests of the Pacific Northwest in America. His exploits, often marred by bad weather and even worse luck, included being attacked by a grizzly bear (he shot it) and losing all his specimens and belongings when caught in a river whirlpool. Think of him next time you see a red-flowering currant (*Ribes sanguineum*) for that was one of 'his' finds. Trees that he brought to Britain included the sugar pine (*Pinus lambertiana*), white pine (*P. strobus*) and ponderosa pine (*P. ponderosa*). 'You will begin to think I manufacture pines at my leisure,' he wrote to his bosses. He is best known for the mighty (yes) Douglas fir (*Pseudotsuga menziesii*), which grows up to 55m (180ft). Douglas was never one for staying home and met his death, aged 35, on an expedition to Hawaii after falling into a pit to capture wild cattle and being gored and trampled by a bullock.

Pseudotsuga menziesii

Pinus strobus

Burle Marx: The Brazilian modernist

Plantsman extraordinaire

Roberto Burle Marx (1909–94) was a man who wore many hats, all with panache. The landscape architect was also a pioneering campaigner of the tropical rainforest. 'The Brazilian forest is now being destroyed across the entire country,' he said in a speech in 1967. 'Everywhere I go, destruction can be seen and felt. This is a state of emergency.' What made his campaigning stance even more remarkable is that this was his first speech as a consular appointee to the dictatorship of General Humberto de Alencar Castelo Branco. The appointment carried its own set of issues, of course, but Burle Marx used it as a platform to raise issues far ahead of their time, including climate change and species extinction, noting the increase of torrential rains and other observable signs.

He was also a painter and artist, a larger-than-life character with a booming voice and love of parties. He designed many public parks and places – including Copacabana Beach promenade – and was in demand worldwide. He created an experimental nursery and stunning artistic garden at his 60-hectare (150-acre) estate in the Rio de Janeiro suburb of Guaratiba, where he loved to entertain the likes of architects Le Corbusier and Frank Lloyd Wright. Burle Marx's botanical collection included 3,500 tropical plants, mostly native to Brazil. He was a plant hunter, embarking on trips throughout Brazil, discovering 37 unidentified species which now carry the Latin name of *burle-marxii*. His garden is a UNESCO World Heritage Site and his words of warning, delivered long ago, still ring true today.

Characteristics of a Burle Marx garden

- Use of tropical plants in borders and islands
- Block-colour planting
- Patterns such as chequerboard lawns or swirling circular shapes
- Bold, sinuous curved lines
- Use of mosaics and sculptural elements
- Innovative water features

Amaryllis brasiliensis

'One may even think of a plant as a note. Played in one chord, it will sound in a particular way; in another chord, its value will be altered. It can be legato, staccato, loud or soft, played on a tuba or on a violin. But it is the same note.'

Roberto Burle Marx

Lancelot 'Capability' Brown: A practical visionary

The hyper-naturalist

Lancelot 'Capability' Brown (1716–83) is one of the most, if not *the* most, famous garden designers of all time, and yet almost everything about his humble early life went against this outcome. He grew up in Northumberland, in far northern England, his father a land agent, his mother a chambermaid. At the age of 16 he became an apprentice gardener at Kirkharle Hall (where his mother worked). At the age of 23, he saddled up and headed south, working at various country houses. His big break came a few years later at the grand country estate of Stowe in Buckinghamshire, when he joined Lord Cobham's staff under the guidance of garden designer William Kent.

Kent was a fan of the natural look, revolutionary in the age of formal gardens. It was a theme that Brown would make his own: he must have exuded confidence and style, for in 1742, at the age of 26, he was appointed head gardener at Stowe. While there, he took on commissions from Lord Cobham's friends, and his hallmark naturalistic style of endless undulating green parkland punctuated with interventions such as lakes and groups of trees became the must-have landscape trend. He was famously hyperactive, visiting estates by horse, sketching out views that would require moving rivers, hills, trees and whatever else was cluttering them.

Brown had a visionary eye – observing a landscape he would quickly assure his patrons of its 'capabilities' – but that was balanced by a down-to-earth practicality. He was, after all, a gardener at heart. Yes, he had a knack for creating a feeling that a landscape was going on infinitely – over the horizon – by changing river courses or creating hills. But he was not above figuring out the nuts and bolts of massive earth-moving schemes and inventing machinery to move large trees. The results, though, were stunningly natural while, of course, being almost brutally unnatural.

Classic signs you are in a Capability Brown landscape

- A grand sweeping drive meticulously planned to impress with different views
- A large serpentine lake mostly surrounded by grass
- Clumps of trees grouped on small hills or in open parkland
- Use of ditches or ha-has (a boundary marker in a ditch) to enable uninterrupted views
- A stone bridge in a classical style (often part of the grand sweeping drive)

Lancelot 'Capability' Brown

'There I make a comma, and there, where a more decided turn is proper, I make a colon. At another part, where an interruption is desirable to break the view, a parenthesis; now a full stop, and then I begin another subject.'

Lancelot Capability Brown, explaining his 'grammatical technique' of designing landscapes, in 1782

Let's just move that village

In 1751, at the age of 35, Capability Brown was asked to redesign the parkland at Croome Court in Worcestershire, which was, by common consent, mostly unproductive marshland. The local village, which was in view of the house, was moved and cloaked with trees. The existing medieval church had to go, being replaced with a new Gothic one. Formal gardens next to the house were removed, replaced by endless grass and parkland, dotted with follies and the like. An almost 2-mile (3-km) long serpentine river was dug by hand and topped with a lake, fed by a system of culverts designed by Brown. In addition, there was a vast variety of plants and a very productive walled garden. Phew! The owners erected a lakeside monument inscribed with the words: 'To the memory of Lancelot Brown, who, by the powers of his inimitable and creative genius, formed this garden scene out of a morass.'

Capability Brown sites to visit

All the following have a significant Brown layer:

1. Alnwick Castle and Hulne Park, Northumberland
2. Ampthill Park, Bedfordshire
3. Ashridge, Hertfordshire
4. Audley End, Essex
5. Basildon Park, Berkshire
6. Berrington Hall, Herefordshire
7. Blenheim Palace, Oxfordshire
8. Bowood House and Gardens, Wiltshire
9. Burton Constable, East Yorkshire
10. Charlecote, Warwickshire
11. Chatsworth, Derbyshire
12. Clandon Park, Surrey
13. Claremont Landscape Garden, Surrey
14. Compton Verney, Warwickshire
15. Coombe Abbey, Warwickshire
16. Croome Court, Worcestershire
17. Gatton Park, Surrey
18. Gibside, Tyne and Wyne
19. Harewood House, West Yorkshire
20. Hatfield Forest, Essex
21. Highclere, Hampshire
22. Ickworth, Suffolk
23. Kirkharle, Northumberland
24. Lacock Abbey, Wiltshire
25. Langley Park, Buckinghamshire
26. Longleat, Wiltshire
27. Luton Hoo, Bedfordshire
28. Milton Abbey, Dorset
29. Moccas Park, Herefordshire
30. Newton House (Dinefwr Castle), Carmarthenshire
31. Petworth, West Sussex
32. Prior Park, Somerset

33. Rothley Lakes, Northumberland
34. Royal Botanic Gardens, Kew, Surrey
35. Sheffield Park, East Sussex
36. Stowe, Buckinghamshire
37. Syon Park, Greater London
38. Temple Newsam, West Yorkshire
39. Trentham Gardens, Staffordshire
40. Uppark, West Sussex
41. Wallington, Northumberland
42. Warwick Castle, Warwickshire
43. Weston Park, Staffordshire
44. Wimpole, Cambridgeshire
45. Woodchester, Gloucestershire
46. Wotton Underwood, Buckinghamshire
47. Wrest Park, Bedfordshire

Derek Jarman: Avant garde-ner

Desert in deepest Dungeness

Born in 1942, Derek Jarman became a filmmaker, artist and activist who, after finding out that he was HIV-positive in December 1986, decided to escape his London life. In 1987 he bought Prospect Cottage, in the fishing hamlet of Dungeness, which he had spotted while filming on the beach with actor Tilda Swinton. It is an otherworldly landscape – flat, with shingle and the sea in front, and a nuclear power plant behind. There are no fences. The sense of surreality is enhanced by the miniature steam train that choo-choos back and forth behind the house. Jarman painted the cottage black with bright yellow windows, and wrote John Donne's poem 'The Sunne Rising' on one side. Then he decided to make his first garden, at the age of 45.

At first, Jarman, who had loved flowers since he was a boy, tried to grow roses. Most failed and he began to work with what grew naturally, including sea kale, and dry Mediterranean plants such as lavender and santolina. At first, he noted, the locals regarded him as someone up to something magical (he liked to wear long Arab robes). He created driftwood sculptures and stone circles, welded metal into various shapes and intermixed these sculptures with fiery poppies. 'Gardening on borrowed time,' he wrote next to one of his planting plans.

Jarman died in 1994 and since then Prospect Cottage has become a place of pilgrimage. His is a garden with lasting power: the sculptures remain amid the windblown and beautiful scene. In 2000, a crowdfunding exercise raised enough money to buy Prospect Cottage for the nation. Jarman said that his garden always attracted a wider audience than his films ever did. 'Gardening is what I should always have done,' he says. 'I shouldn't have become involved in cinema. That's for fools.'

Plant in the style of Prospect Cottage

- Santolina
- Red valerian (*Centranthus ruber*)
- Sea kale (*Crambe maritima*)
- Lavender
- Poppies

'My garden's boundaries are the horizon.'

Derek Jarman

Santolina chamaecyparissus

Lavandula angustifolia

Crambe maritima

Other ground-breaking gardeners

Frederick Law Olmsted (1822–1903)

American landscape architect and park-maker Olmsted, with his partner, Calvert Vaux, designed Central Park in New York City as well as many others, including Prospect Park in Brooklyn. He was also a journalist and a pioneering conservationist.

Karl Foerster (1874–1970)

Foerster, whose nursery was in Potsdam-Bornim in Germany, was known for his selection of tough, low-maintenance, hardy perennials that were beautiful, resilient and endurable. He is best known for a grass that, as legend has it, he saw from a train window and, after pulling the emergency brake, collected. *Calamagrostis* x *acutiflora* 'Karl Foerster' is now one of the most popular ornamental grasses in the world.

Olive Muriel Pink (1884–1975) and Johnny Jambijimba Yannarilyi (d. 1973)

Botanical artist Olive Pink and Walpiri man Johnny Jambijimba Yannarilyi worked together to transform a piece of unoccupied land, grazed by feral goats and rabbits, in Alice Springs, Australia, into a showcase for native plants. The botanical garden, which opened in 1986, contains 600 arid-zone plant species from central Australia, including many that are threatened.

Calamagrostis epigejos

Edna Walling (1895–1973)

Australian landscape designer, conservationist, writer and photographer Walling is credited with steering away from an Anglocentric style towards one that chimed with Australia's climate and landscapes. She drew inspiration from the Australian bush, with a naturalistic style using boulders, rocky outcrops and native plants.

Edavalath Kakkat Janaki Ammal (1897–1984)

EK Janaki Ammal was a pioneering botanist and plant cytologist who is often called 'India's first female botanist'. Her studies on sugarcane developed hybrids that helped India to grow its own supply. Her legacy includes a successful campaign to protect Silent Valley National Park in Kerala, with its many rare orchids, and she worked at RHS Garden Wisley, in Surrey, England, UK, from 1946–51.

Thomas Church (1902–78)

Church was an American landscape architect known for using Modernist principles in garden design and being a pioneer of the 'California style'. His four key principles, as detailed in his 1955 book *Gardens Are For People*, are: unity, function, simplicity and scale. 'A garden should have no beginning and no end and should be pleasing when seen from any angle, not only from the house,' he wrote.

Luis Barragán (1902–88)

Renowned Mexican architect and engineer Barragán is famous for his use of block colours and forms in his buildings, and for a philosophy in which the garden was equally as important as the building. 'The perfect garden, no matter its size, should enclose nothing less than the entire Universe,' he said. His home and garden in Mexico City is a UNESCO World Heritage Site.

Piet Oudolf (1944–)

Oudolf, the most successful of the Dutch New Wave founders, has established a global reputation for the New Perennial movement. His intricate designs, which include the High Line in New York City (see page 154), focus on naturalistic planting as well as biodiversity. He is a man who likes to break the rules, and his personal garden, in Hummelo, The Netherlands, was a place of pilgrimage until it closed to the public in 2018.

Ron Finley

The fashion designer and self-styled Gangsta Gardener, who doesn't give his age, has started a horticultural revolution in the asphalt city that is South Central Los Angeles. He began by planting a food forest on the city-owned strip of land outside his home. 'It starts with one seed,' he said. One of his campaigns led to no more fines being levied for kerbside gardens in LA.

No-dig pioneers

There's no work . . .

American author Ruth Stout (1884–1980) called her method of gardening 'no work', which, needless to say, made it rather popular. Her innovative methods were born out of necessity. When she began to garden at the family farm in Poverty Hollow, near Reading, Connecticut, in the 1940s, she had to wait for someone to come to plough the soil. Often that someone was delayed and, after a while, Stout decided she couldn't wait any longer. In 1944 she decided to give up on ploughing and began improving her soil with a mulch that was at least 20cm (8in) deep. She wasn't fussy about what was in her mulch either: basically anything good enough for the compost heap was good enough for the mulch. Her best-known book is the succinctly titled *Gardening Without Work* (1961).

. . . And no digging

Esther Deans (1911–2008) was a writer, gardener and conservationist who, at one point, became Australia's most published author. Her bestseller was called *Esther Deans' Gardening Book: Growing Without Digging* (1977) and she became famous in the 1970s and 1980s for exactly that. Her method was born out of what could be called desperation, as she faced working with soil in her back garden in Sydney that was pure clay. Her solution was to abandon the idea of digging down and, instead, to build the garden up, layer by layer. The layers were made of hay, compost, paper, chicken manure, straw and fertiliser. She continued gardening until she was in her nineties. She once commented: 'Whatever I do, I apply myself, whether I like it or not. And I think that is a number one lesson we should all learn.'

Cucurbita maxima

'I planted the seeds, and it was just like something magic. The zucchini just grew and grew and the beans just grew. It was incredible what happened.'

Esther Deans

Journey to the Far East

Zheng Guogu and the Age of Empires

Liao Garden in Yangjiang, Canton, on the southwest coast of China, is a garden, a living work of art and a studio combined. It belongs to artist Zheng Guogu, born in 1970, who began in 2000 by calling the garden the Age of Empire, after the computer game that he was obsessed with at the time. 'There were about 80 million people playing Age of Empires in China at that time, so I thought to import the game's elements into the idea of the Chinese garden but not the kind of Ming and Qing, or Suzhou gardens or later the Li Garden in Kaping, but a contemporary one. So we made a garden of THIS era.' He and his team set to work, digging canals and creating water features and buildings including a golden pyramid. There is also 'The Museum of the Wind', which is a series of open layers connected by staircases and pathways. The garden, now called Liao Garden, continues to evolve with an atrium, a frangipani garden (*Plumeria* is a plant not often seen in Chinese gardens) and a feature called a *tianchi*, with contours curved like terraced fields, and each of the more than 12 layers has water like a pool.

Plumeria rubra

'We proposed to make a museum but we had nothing to exhibit! The museum only exhibits the wind, which is different every day.'

Zheng Guogu

Tokachi: Vision of the future

The Tokachi Millennium Forest was the brainchild of media magnate Mitsushige Hayashi. He bought the 400 hectares (1,000 acres) of land on Japan's northern-most island, Hokkaido, in the 1990s with the idea of offsetting the carbon footprint of his newspaper business. He wanted to create a conservation project with a thousand-year sustainable vision. The two people who provided the practical impetus to create this vision on land are British garden designer Dan Pearson and Japanese horticulturist Midori Shintani. In the plateaus and wooded foothills of the Hidaka Mountains they have pioneered a new way of gardening.

It's a complete team effort. Pearson, a garden superstar in Britain with a painterly naturalistic approach, has been working on the project since 2000 and Shintani since 2007. She brings an intuitive way of working which is imbued with a sensitivity to Japanese traditions and ways. Pearson's bold meadow, enlivened by an undulating landscape created out of what was a flat field, forms part of what he calls the 'garden layer' to the park. Swathes of perennials and grasses billow here, the colours changing constantly, a majestically beautiful landscape. Both the gardens and forests yield produce.

Their approach is epitomized by the word *satoyama*: this is both a geographical term denoting the area between the foothills and agricultural land, and an ethos that is all about people living harmoniously within this landscape. Balance is crucial, and a core belief is that people should only take what they need from the land. For Pearson, thoughtful integration has been the key.

Shintani sees tradition, such as Japanese nature worship, as important and something that can be developed into something modern. Nature provides signs of when it is time to act: the first call of the cuckoo means that the snow will go soon and that it's time to plant seeds. She references the Japanese nature calendar with its 72 micro-seasons. This calendar is both poetic and practical in its guidance, which fits perfectly at this place.

Iris ensata

Poetry in motion: 72 micro-seasons

In Japan there are not just four seasons of spring, summer, autumn and winter: instead there are 24 major divisions – each with a short heading – which are divided further into 72 micro-seasons of about five days each. The original idea came from China but was rewritten in the 17th century to fit Japan's climate. Dates can vary, according to calendar year. It provides a poetic guide even though, obviously, sometimes reality doesn't always oblige.

Beginning of spring
East wind melts the ice (4–8 Feb)
Bush warblers start singing in the mountains (9–13 Feb)
Fish emerge from the ice (14–18 Feb)

Rainwater
Rain moistens the soil (19–23 Feb)
Mist begins to linger (24–28 Feb)
Trees bud, grass sprouts (1–5 March)

Insects awaken
Hibernating insects surface (6–10 March)
First peach blossoms (11–15 March)
Caterpillars become butterflies (16–20 March)

Spring equinox
Sparrows start to nest (21–25 March)
First cherry blossoms (26–30 March)
Distant thunder (31 March–4 April)

Pure and clear
Swallows return (5–9 April)
Geese fly north (10–14 April)
First rainbows (15–19 April)

Grain rains
First reeds sprout (20–24 April)
Last frost, rice seedlings grow (25–29 April)
Peonies bloom (30 April–4 May)

Summer begins
Frogs start singing (5–9 May)
Worms surface (10–14 May)
Bamboo shoot sprouts (15–20 May)

Lesser ripening
Silkworms start feasting on mulberry leaves (20–25 May)
Safflowers bloom (26–30 May)
Wheat ripens (31 May–5 June)

Grain beards and seeds
Praying mantises hatch (6–10 June)
Rotten grass becomes fireflies (11–15 June)
Plums turn yellow (16–20 June)

Summer solstice
Self-heal withers (21–26 June)
Irises bloom (27 June–1 July)
Crow dipper sprouts (2–6 July)

Lesser heat
Warm winds blow (7–11 July)
First lotus blossoms (12–16 July)
Hawks learn to fly (17–22 July)

Greater heat
Paulownia trees have seed (23–28 July)
Earth is damp, air is humid (29 July–2 Aug)
Great rains sometimes fall (3–7 Aug)

Beginning of autumn
Cool winds blow (8–12 Aug)
Evening cicadas sing (13–17 Aug)
Thick fog descends (18–22 Aug)

Manageable heat
Cotton flowers bloom (23–27 Aug)
Heat starts to die down (28 Aug–1 Sept)
Rice ripens (2–7 Sept)

White dew
Dew glistens white on grass (8–12 Sept)
Wagtails sing (13–17 Sept)
Swallows leave (18–22 Sept)

Autumn equinox
Thunder ceases (23–27 Sept)
Insects hole up underground (28 Sept–2 Oct)
Farmers drain fields (3–7 Oct)

Cold dew
Wild geese return (8–12 Oct)
Chrysanthemums bloom (13–17 Oct)
Crickets chirp around the door (18–22 Oct)

Frost falls
First frost (23–27 Oct)
Light rains sometimes fall (28 Oct–1 Nov)
Maple leaves and ivy turn yellow (2–6 Nov)

Winter begins
Camellias bloom (7–11 Nov)
Land starts to freeze (12–16 Nov)
Daffodils bloom (17–21 Nov)

Lesser snow
Rainbows hide (22–26 Nov)
North wind blows the leaves from the trees (27 Nov–1 Dec)
Tachibana citrus tree leaves start to turn yellow (2–6 Dec)

Greater snow
Cold sets in, winter begins (7–11 Dec)
Bears start hibernating (12–16 Dec)
Salmon gather and swim upstream (17–21 Dec)

Winter solstice
Self-heal sprouts (22–26 Dec)
Deer shed antlers (27–31 Dec)
Wheat sprouts under snow (1–4 Jan)

Lesser cold
Parsley flourishes (5–9 Jan)
Springs thaw (10–14 Jan)
Pheasants start to call (15–19 Jan)

Greater cold
Butterburs bud (20–24 Jan)
Ice thickens on streams (25–29 Jan)
Hens start laying eggs (30 Jan–2 Feb)

Famous folk on gardening

'Until you dig a hole, you plant a tree, you water it and make it survive, you haven't done a thing. You are just talking.'

Wangari Maathai, Kenyan political activist and Nobel Prize winner (1940–2011)

'The best time to plant a tree was 20 years ago. The next best is today.'

Chinese proverb

'Flowers are restful to look at. They have neither conflict or emotions.'

Sigmund Freud, Austrian psychiatrist (1856–1939)

'There is nothing that is comparable to it, as satisfactory or as thrilling as gathering vegetables one has grown.'

Alice B. Toklas, American writer (1877–1967)

'We can complain because rose bushes have thorns or rejoice because thorn bushes have roses.'

Jean-Baptiste Alphonse Karr, French writer (1808–90), though often attributed to Abraham Lincoln

Rosa × alba

'The best place to find God is in a garden. You can dig for him there.'

George Bernard Shaw, Irish playwright (1856–1950)

Vitis vinifera

'Gardening is the work of a lifetime: you never finish.'

Oscar de la Renta, Dominican fashion designer (1932–2014)

'My life now is just trees. Trees and champagne.'

Dame Judi Dench, English actor (b. 1934)

'To forget how to dig the earth and to tend the soil is to forget ourselves.'

Mahatma Gandhi, Indian politician (1869–1948)

'To plant a garden is to dream of tomorrow.'

Audrey Hepburn, British actor (1929–93)

'In the spring, at the end of the day, you should smell like dirt.'

Margaret Atwood, Canadian writer (b. 1939)

'What is a weed? A plant whose virtues have not been discovered.'

Ralph Waldo Emerson, American philosopher (1803–82)

Ribes nigrum

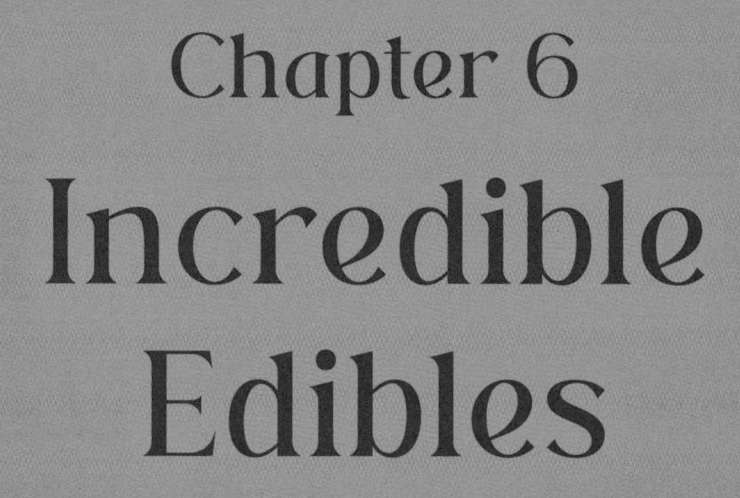

Dive into a cornucopia of information that celebrates how and what our gardens grow. It's not all about food and drink, though there's plenty of that, with recipes and tips on everything from the food of love to the drinks of the gods.

Alcoholic plants

It all begins with photosynthesis

At the root of every great drink, there lies a plant. Plants take in sunlight and carbon dioxide and convert it to oxygen and – this is the key – sugar. When you add yeast to sugar, alcohol is born. Different plants create an equally diverse array of alcoholic drinks. Sugar cane leads to rum, agave to tequila, grapes to wine, potato to vodka, rice to sake and barley or wheat to beer, whisky, (more) vodka and gin. There are even more exotic botanical concoctions out there, though, including a ceremonial drink called *mudai*, made in Chile from the seeds of the monkey puzzle tree (*Araucaria auraucana*).

Artemesia absinthium

Mad, bad but not dangerous to grow

Wormwood (*Artemesia absinthium*), with its feathery grey-green foliage and distinctive bitter scent, is a common herb that has many uses as an insecticide and companion plant, as well as providing a colour contrast in floral bouquets. But it also hides a history of madness and addiction. It is one of the main ingredients, along with green anise and sweet fennel, of absinthe, also known as the 'green fairy', an alcoholic (up to 80%ABV) concoction with a notorious reputation for turning its devotees into glassy-eyed benumbed addicts. Just look at the vacant stare of the young woman in Degas' 1876 painting *L'Absinthe*. The drink, a favourite of creatives including Toulouse-Lautrec and Ernest Hemingway, was banned in the USA in 1912 and France in 1914, but now is back in the cabinet.

Make a blackberry wine

This makes a medium dry wine. You'll need a large bowl, two gallon (4L) jars or jugs, a straining bag, a cotton bung and an air lock. Everything should be sterilized before use.

Rubus allegheniensis

- 1.5kg (3lb 5oz) blackberries
- 2 teaspoons active dried yeast
- 1kg (2lb 4oz) sugar
- 1 lemon
- 1 orange

1. Pour 2.2 litres (9 cups) of boiling water on the blackberries in a large bowl and mash with a spoon or rolling pin, or (when cooler) by hand.
2. Leave for three days, stirring daily.
3. Strain and reserve the juice.
4. Add the sugar to 1 litre (4 cups) of water and boil until dissolved.
5. Pour the juice and sugar syrup into the gallon jar or jug.
6. Zest and juice the lemon and orange and add to the jar.
7. Add cool (but previously boiled) water to make up the gallon.
8. Add the yeast.
9. Use a cotton 'bung' and leave for a few days.
10. Fit an airlock and leave to ferment for three months.
11. Decant into another gallon jar and leave for another six months, when it should be ready for bottling and drinking.

Recipe adapted from *Wild Food* (2014) by Roger Phillips, who advised that he needed a little pectinol to clear the wine.

Borago officinalis

Rosa rubrifolia

Grow a cocktail garnish

Plant lavender, rosemary, thyme, cucumber, lime, basil, nasturtiums, violas and/or borage. Recipe: Pick your flowers, add booze, enjoy.

Tap a birch or sycamore for sap

1. Find a mature tree during early spring.
2. Drill a hole, sloping upwards, about 45cm (18in) from the ground.
3. Insert a tube, the other end of which leads to a collecting bottle.
4. Cork up the hole when finished, to stop your tree from bleeding out.
5. Head home to make wine or beer.

Forage a glass of wine

Here's just some of the many fruits and flowers found in hedgerows and the countryside that can be used to make wine:

- Blackberries
- Elderberries
- Haws
- Oak leaves
- Rosehips
- Gorse flowers
- Red clover
- Dandelion flowers
- Broom flowers
- Birch sap

Gin: a tipsy bouquet in a bottle

Of all the spirits out there, perhaps the most botanically spirited of them all is gin. The alcohol, from grain such as barley or wheat, is just the start of it. Common juniper berries (*Juniperus communis*) give gin its distinctive flavour and are required by law in the EU to be the core botanical of any drink called gin. After that, it's like some sort of botanical bottle party. Botanicals used in gin include coriander seed, citrus peel, lavender, bay leaf, cubeb (*Piper cubeba*) and ginger. Then, of course, comes the extra flavourings added at the bar, which can include borage, rose petals and mint.

Juniperus virginiana

No, you can't grow your own Pimm's

James Pimm was born in 1798, the son of a yeoman farmer from north Kent, England, UK. By his twenties, he gave his occupation as an 'oyster dealer' (Kent being the main source of oysters) and soon this expanded to owning an oyster bar in the City of London. 'Pimms in the Poultry' served oysters and lobsters, typically washed down with either stout or rum-based 'house cup'. But Pimm was more adventuresome in the 1840s, creating a drink (Pimm's No. 1) using gin (then all the rage) mixed with fruit extract and herbs. The recipe is still a secret, though some of the botanicals include bitter orange, lemon, juniper and ginger.

Citrus × aurantium

High and low: mountain and desert

Have a prickly pear (without the prickles)

Cacti may look a little spiky, but they can be delicious. The wide flat pads, or *nopales*, of the prickly pear (*Opuntia ficus-indica*) have a lemony flavour with a bit of a crunch, and are used in many dishes including salads and eggs and as a filling. There are also the small oval fruits, which grow on the end of the pads, which, depending on ripeness, range from slightly sweet to positively syrupy. When picking the fruits or the pads, be careful as they are covered in visible thorns or spines as well as tiny, barbed hairs called glochids. These will already have been removed if you buy them in a market but, if not, you will need to remove them, wearing gloves, by using a knife to pare the fruit or, with the pads, removing the edges and skimming over the surface.

Make mine an alpine

The hills are, indeed, alive, and not just with the sound of music. Wild edibles on a hilly or mountain hike can include alpine sorrel (*Oxyria digyna*), alpine chives and, sweetly, alpine strawberries (*Fragaria vesca*) – plenty of edibles for an alpine salad. Then there's edelweiss (*Leontopodium alpinum* ssp. *nivale*), rose bay willowherb (*Chamaenerion angustifolium*), goat's beard (*Tragopogon pratensis*), thyme and wood sorrel, to name a few.

Oxalis acetosella

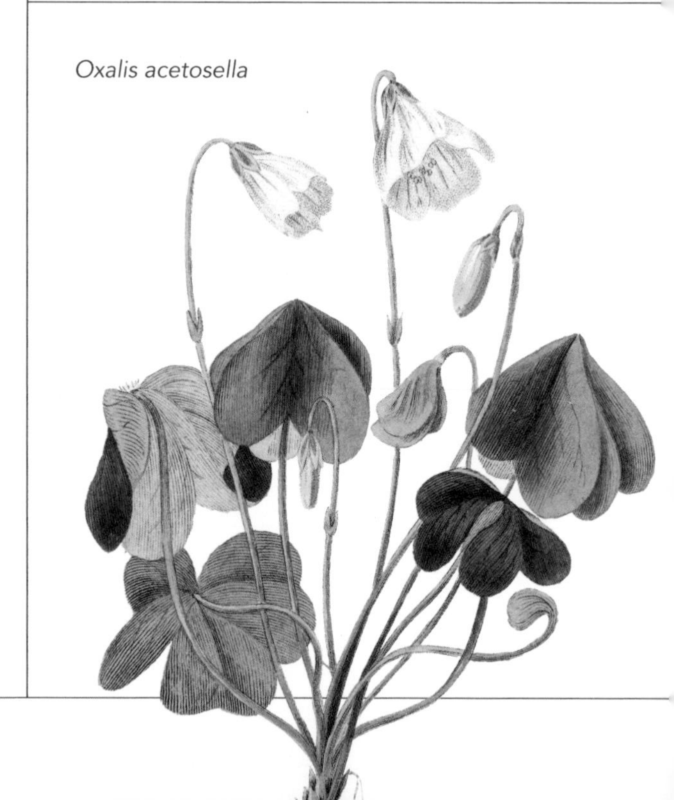

Have you tried . . . ?

Mesquite (*Prosopis* spp.): Mostly used as a smoky flavouring for meats or as an alternative to brown sugar, but its pods, which look like garden peas, can be ground into a flour.

Pinyon pine (*Pinus monyphylla*): This is one of the species pine nuts come from, but the trees must be between 10 and 25 years old before they produce them.

Yucca (*Yucca* spp.): The fruit of thick-leaved varieties can be eaten raw or even grilled.

Tamarisk (*Tamarix* spp.): The leaves and young sheets of this shrub/tree can be eaten raw or cooked.

Doum palm (*Hyphaene thebaica*): This palm has hard brown fruits that look like small coconuts and can be eaten raw or cooked. The leaves can be used to make tea or as a seasoning.

Bramble acacia (*Acacia victoriae*): These trees grow widely in Australia. The seeds, which have a nutty taste, can be roasted, ground and added to flour as well as to ice cream and coffee.

Old man saltbush (*Atriplex nummularia*): This is another species from Australia. It is likely only salty when grown on saline soils. The leaves are used like spinach, and the seeds are roasted.

Yucca gloriosa

Tamarix gallica

Wild foraging

Go wild in your garden

Why not go foraging in your own back garden? It's easy enough to do. Almost all 'wild' plants are now also sold in garden centres or online. Plant your favourites and then, when the time is right, go out and take your pick of them all.

Neighbourhood flavour map

What grows in your neighbourhood that is wild and edible? It's one of those questions that, when you start to answer it, makes you realize there is always something new to learn. You can take inspiration from the likes of a chef named John Evans Ravenna, who, more than a decade ago, had a restaurant in Peru's capital Lima, which is on the coast and has a population of 11 million (and counting). As one report noted, Lima (like so many cities) is supposed to be a place where nothing grows. Evans had other ideas: he went about creating a 'flavour map' of the city's Costa Verde. Where others saw concrete and weeds, he found so much more: the street tree *Schinus molle* with its pink peppercorns; sea lettuce and algae among the coastal rocks, and a type of goosefoot called *cañihua* (*Chenopodium pallidicaule*), which is similar to quinoa. What is exciting is that your flavour map won't be like anyone else's. It's a fulfilling – and potentially tasty – exercise.

Traits of a champion forager

You do not need to live in the countryside – or even 'ruburbia', as it is sometimes called, that area between the suburbs and open country – to forage. You will need a sharp eye for greenery, a curiosity about wild edibles, an identification guide and a desire to research how your list of plants can be turned into seasonings, soups and a host of other foods. Every place, not to mention country, is different.

Hedgerow heaven

There is such a thing as 'hedgerow jelly', but what kind it is depends on the hedge. There may be hawthorn, elderberry or the abundant hips of a wild rose. In Australia, there is lilly-pilly (*Syzygium smithii*), with its bright berries. Whatever the berry, the key to success is to make sure you have enough pectin. Lilly-pilly, for instance, should have enough pectin on its own but, for the likes of haw or rosehip jelly, you will need to add an equal weight of crab apples or other edible apple. This recipe, adapted from Christine Iverson's *The Hedgerow Apothecary* (2019), is for a variety of hedgerow fruits.

Syzygium smithii

- 1kg (2lb 4oz) of crab apples or eating apples
- 1kg (2lb 4oz) mixed foraged hedgerow fruits
- Sugar

1. Clean out the stalks and cut up the apples and hedgerow fruits (chop up rosehips in particular).

2. Put all in a large pan and add enough water so the fruit floats. Simmer for 30–45 minutes.

3. Pour into a colander lined with muslin and let the juice drip into a pan. Leave overnight.

4. Measure the juice and, for every 600ml (20fl oz), add perhaps 450g (1lb) of sugar (some like it sweeter).

5. Bring slowly to the boil and then let it bubble away until 'setting point' has been reached. (You'll know this through the 'wrinkle test': freeze a plate, put a dab of jelly on it and leave for a moment or two, then poke it with your finger. If it wrinkles, it's ready; if it is still liquid, keep the pan on the boil.)

6. Pour into sterilized jars and label.

Aronia
arbutifolia
Ribes
uva-crispa

A forager's kit

- Pocket-sized field guide and/or identification plant app
- Gloves, long sleeves and trousers – to protect against brambles, thorns and nettles (or, in dry landscapes, cacti)
- Secateurs are best, sturdy scissors can work too
- Bags or a basket
- Map or paper and pen, to note where you find certain plants

Morus nigra

Foraging etiquette

- If you want to gain access to private land you will need the permission of the land owner. You can always offer to share your foraging spoils with them.
- Avoid all sites that have been set aside specifically for scientific interest and/or wildlife.
- 'Do no harm' is the motto, and that means avoiding nesting areas.
- Take only what you need and leave plenty for wildlife.
- Do not pick or take anything that is endangered, and never dig up plants or bulbs.
- Avoid areas next to fields that may have been sprayed with pesticides or herbicides, or busy, polluted roads.

Worldwide wild berries

All of these berries grow wild in various parts of the world and can be used to make jellies, jams, vinegars, syrups and sauces:

- Barberries
- Blackberries
- Blueberries
- Buffaloberries
- Chokeberries
- Cloudberries
- Dewberries
- Elderberries
- Fuchsia
- Gooseberries
- Huckleberries
- Mulberries
- Muscadine (wild grapes)
- Raspberries
- Rowan
- Salmonberries
- Saskatoons
- Sloes
- Strawberries

Famous fruit trees

What was the 'forbidden fruit'?

It was the fruit that started it all: Eve ate it, shared it with Adam, then they suddenly realized they were naked. They were thrown out of paradise and, well, you know the story. For some time, the apple is often seen as the culprit, but that seems highly unlikely, not least because the apple originates from Central Asia, which is quite a way from the Garden of Eden, widely believed to have been in the Middle East.

So here are some options. Fig is often mentioned, and we know there was one growing nearby because Adam and Eve used its leaves to cover themselves. But sometimes the skin of a fig can be a little tough. No one mentions the idea of peeling the forbidden fruit, but we can't rule it out. Other options could be a citron or pomegranate (though peeling really would be an issue) or perhaps an apricot or peach.

Ficus carica

The Franken tree

This tree wasn't around during Genesis, which is a pity, for it offers such an array of options. It is made up of up to 40 different types of fruit, creating what amounts to the ultimate fruit-salad combo. Come spring, it bursts into a sequential blaze of variously coloured blossom – pink, crimson, fuchsia, red, white – which is followed by a variety of stone fruits. The idea was thought up by Sam Van Aken, an art professor at the University of Syracuse in New York State, USA. He grew up on a farm and had seen grafting as a child. 'It is like Dr Seuss and Frankenstein and all those amazing things you think of when you're a kid.'

Van Aken sees his 'Tree of 40 Fruit' as a living work of art. They are not particularly easy to create, as his 'art nursery' on campus shows. He works with 250 varieties of stone fruit trees from the *Prunus* genus, including almonds, peaches, plums, apricots, cherries and nectarines, most of which are heirloom or unusual varieties. The rootstock (plum) needs to be three years old with four or five branches before he starts (bud) grafting on other types of fruit, usually several a year. He has a series of trees on the go at any one time before planting them out. But are these trees art or, more practically, just a very complicated dessert?

Prunus domestica

Prunus cerasus

King of fruits

The royal seal of approval

Mashed into a rather delicious sauce by the Carib people on the island of Guadeloupe, pineapples have had a royal reputation in Europe since they were 'discovered' in 1493 by Christopher Columbus. By the time he got back to Spain, only one of his pineapples was unrotten, but that was enough to impress King Ferdinand, who pronounced the fruit 'superior' to all others. Pineapples became a status symbol, so valuable that, outside of royal households, they were often rented to act as dinner-party centrepieces.

What do pineapples have in common with the pyramids?

Both pineapples and the pyramids of Giza are illustrations of what the ancient Greeks called the Divine Proportion, or Golden Ratio. This is *phi*, which can be illustrated by the Fibonacci sequence, named after Italian mathematician Fibonacci, also known as Leonardo of Pisa. On every pineapple, there are 5, 8 or 13 spirals (all Fibonacci numbers), depending on which slope you measure. The leaves are also placed using the Golden Ratio.

Pineapples in architecture

Pineapples, a symbol in Europe of wealth and status as well as hospitality since the Renaissance, became a favourite motif for architects. (They also made a change from acorns.) You'll see them on rooftops, railings, doors and entranceways. Some of the most spectacular are:

- The 1761 summerhouse in Dunmore Park, Stirlingshire, Scotland, UK, has a 14m (46ft) stone cupola, carved as a pineapple as its central feature.
- Sir Christopher Wren placed a golden pineapple on top the western towers at St Paul's Cathedral, London, England, UK.
- Sir John Soane put a pineapple on top of the mausoleum he designed (which entombs his wife, son and Sir John himself), in Old St Pancras Churchyard, London, England, UK.

Hothouses of Holland

The European craze for pineapples was only heightened by the fact they proved so maddeningly difficult to grow outside the Tropics. The first requirement was a hothouse. In Europe, it was the Dutch with their extensive colonial trade links who succeeded first. It was a woman who grew the first pineapple in Europe: the wealthy, well-educated (and ambitious) Agneta Block accomplished this in 1687 in her specially constructed hothouses in Vijverhof. Needless to say, the arrival of William of Orange in England in 1688 was good news for pineapple lovers: after he became king he was followed to Hampton Court by a retinue of gardeners with expertise in pineapple hothousing.

Don't pollinate the pineapples!

Pineapples are perennials that grow all year round. They are bromeliads and, although not epiphytic (growing on other plants), they do share some attributes, such as the ability to store water in their leaves. Their small, purple, tubular flowers are grouped together on a central stalk. These are asexual plants that can self-pollinate, though, in the wild, they are also pollinated by bats and hummingbirds, resulting in a plethora of seeds. One of the key goals of pineapple cultivation was to stop this seedy situation: in Hawaii, a mega-producer of pineapples during the 20th century, the import of hummingbirds was banned.

Trochilus furcatus

Pineapple diet, anyone?

The pineapple is the only known source of bromelain, an enzyme that digests protein. It is a main ingredient of meat tenderizer, and explains why pineapple is so often served with meats such as ham. There is a certain flesh-eating theme here, as it was first cultivated in about 2000 BCE by the cannibalistic Tupi-Guarani Indians in Amazonia. The proteolytic enzyme, which aids digestion and supposedly targets belly fat, also explains why pineapples are so popular in diets, including the *Sexy Pineapple Diet*, published by Danish psychologist Sten Hegeler and his wife Inge in 1970.

Ananas comosus

Slow-growing home-grown pineapple

What you'll need:

- A ripe pineapple
- A pot
- Compost
- Lots of patience

1. Cut or twist off the crown just above the fruit and let dry for several days.

2. Peel away three or four layers of the lower leaves. You may see the first roots appearing (if not, a circle of brown dots indicates where they will be).

3. You can put it into a jar filled with water to encourage roots before planting up or just place directly into a pot filled with moist compost.

4. Find a sunny, protected position. Do not allow to become chilled.

5. Wait two years.

Legend of the Three Sisters

Stronger together

The Three Sisters traditional method of planting can be traced back many hundreds of years to the Indigenous American tribes of the Haudenosaunee Confederacy, also commonly known as the Iroquois Confederacy, who lived in upstate New York, spreading to Lake Ontario in what is now the USA and Canada. The confederation includes the Mohawk, Oneida, Onondaga, Cayuga and Seneca tribes. The Three Sisters also spread to other tribes.

For the Haudenosaunee, the Three Sisters are not just corn, beans and squash but also divine gifts. In Seneca legend, they are known as *Diohako*, Sustainers of Life. The sisters feature in the Iroquois creation myth as well as various other legends. These usually feature three sisters who were very different in looks and temperament – one tall with long silky hair, one small and muscular, another who was of average height but giving in nature. The legends centre on how the girls loved one another but became separated then reunited, stronger together, forever. The plants are seen as sacred, providing for everyone, and ensuring the survival of the people.

The traditional method

The planting was done by the women, who were responsible for farming, and it began by placing several kernels of corn in a hole. As the seedlings grew (at least 30cm/1ft high), the soil was mounded around them. The final mound was about 30cm (1ft) high and 60cm (2ft) wide. The rows were about one step apart. Two or three weeks after the corn was planted, seeds for climbing beans were planted in the same mounds. Between the rows, seeds for a low-growing crop such as squash (or pumpkins) was planted, up to a week or two later.

Why it works: The cornstalks serve as bean poles and the beans add nitrogen to the soil. The low-growing squash, with its large leaves and long vines, shade the ground, preserving moisture and inhibiting weeds.

Celebrating corn

Corn was especially important to Indigenous Americans because it had so many uses. This was white corn (*Zea mays*), which is not as sweet as the yellow type (*Z. mays* subsp. *mays*) more commonly grown today. The corn was pounded into meal and made into bread, hominy and pudding. It was made into soup and *succotash*, a stew that also contained beans and squash. Corn husks were woven into mats and baskets; the cobs used as scrubbers. The start of the harvest was celebrated every year, in August or early September, with a Green Corn Ceremony.

Cucurbita maxima

Zea mays

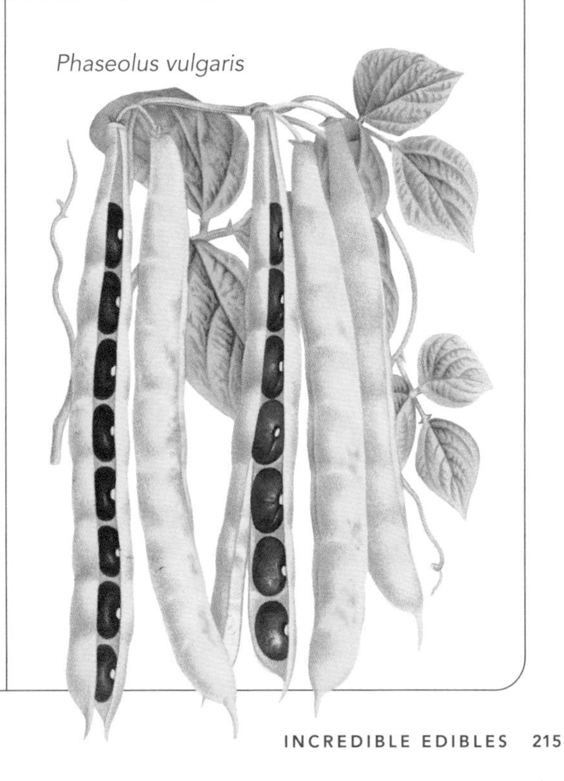

Phaseolus vulgaris

Humongous veg

Grow a huge pumpkin

A pumpkin is a member of the Cucurbitaceae family, along with cucumbers and squashes. Pumpkins, which are known as vegetables but are technically fruits, are normally the size of a basketball. So how big *is* the largest pumpkin? The record, which seems to be broken every year, hovers around the 1,224kg (2,700lb) mark: this is a pumpkin the size of a small car. Now, that's an awful lot of pumpkin for soup, stews and pies. And, it must be said, an extremely impressive Jack O'Lantern. Here's how to go about it:

Choose the right seed: The clue is in the name, and varieties include 'Prizewinner' and 'Big Moose'. But the one with the best track record is 'Atlantic Giant'. Remember that, as you are growing a giant, you will have plenty of seeds of your own by the end of the season.

You'll need some sun: Pumpkins love sun, and giant ones love even more rays, so find the sunniest spot that you have that is also not overly exposed to wind.

Create the perfect bed: Make sure you allow enough space for your giant(s) and, in spring or, even better, autumn, dig a large hole and fill it with well-rotted manure and compost. Or you could think about growing the pumpkin on your compost heap. Sprinkle organic fertiliser such as blood, fish and bone. Pumpkins like neutral or slightly acidic soil.

Plant your seed: If there is a chance of frost early spring, then start seeds off inside in a small pot filled with organic peat-free seed compost. The outer skin of giant seeds can be tough, and some people like to 'chit' them, filing them a bit (only the sides, not the pointy end) with an emery board. Or chit a few and plant a few, and see what happens. Keep moist but don't overwater. When roots reach the bottom of the pot, transplant to a larger one or, if warm enough outside, put outside under cover.

Transplant: In cold climates, you may have to wait until late spring to transplant outside. Make sure your planting hole is deep enough to accommodate all the roots. Sprinkle water on the roots (not the leaves). Don't overwater, and you might want to leave it for a few days as there will be moisture in the soil. For a giant, you will need to fertilise weekly.

'I'm often asked, why giant vegetables? The truth is, they're fun.'

Charlie McCormack,
English giant vegetable grower

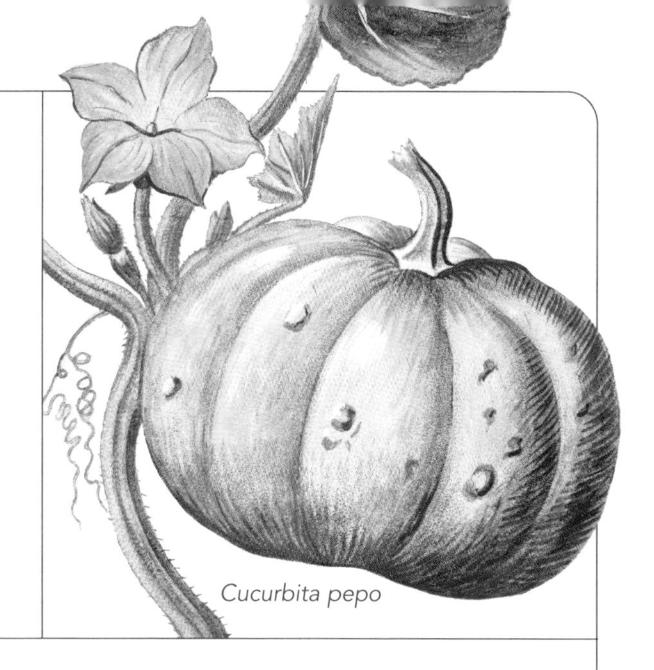

Cucurbita pepo

Pollination: The male flowers come first, then the females. The male is smaller and the female is recognizable because, at the base, right underneath the flower, there is a mini-fruit. Bees and natural pollinators should be fine, though some growers do hand-pollinate.

Prepare for greatness: Remember that the vine is going to be massive. It needs to be strong and well established before you allow the female flowers to bear fruit. Be on patrol once the vine reaches 6m (20ft), and pinch out the sideshoots and tips. Allow three or perhaps four flowers (spaced out) to produce fruit. You may want to keep all of them, but if you want a whopper, then once they reach volleyball size, you'll need to cull all but one. This is an only child that is going to want it all, and remember that means keeping the vine and roots healthy too.

Cardboard to the rescue: Place your pumpkin(s) on a thick piece of cardboard or a piece of plywood to avoid rotting and damage from soil-dwelling insects.

Watch it grow: By mid-summer, the pumpkin will be expanding like mad: night-time is the right time for pumpkins to grow, and some expand 13cm (5in) in circumference every night.

TLC: How much time and effort you devote to your giant depends on you, of course, but you will need to make sure it has enough water (but not too much) and is protected from extreme weather. Some growers cover their pumpkins before heavy rains or use shade tents on hot days. Everyone has their special tips, including the idea that the key to success lies in peeing in a bucket every morning and then passing it on. Not mandatory, or even scientifically proven.

Ingenious urban farms

Is it a car park or a farm?

The largest urban farm in Paris, France, looks like an underground car park because, well, it used to be exactly that. It is underneath a nondescript 1970s social housing building in the unfashionable part of the northern 18th arrondissement. For a time, the 9,000sq m (9,700sq ft) of concrete was abandoned, haunted by squatters and drug users, but, in 2017, it was cleared out. In came a company called Cycloponics, which wanted to grow mushrooms underground. The organic farm, called La Caverne, does indeed grow mushrooms galore, and some parking bays have been converted for herbs and salad greens. There's also a good crop of endive, sometimes called the goths of the vegetable world, which likes to grow in the dark. All are delivered to restaurants – by bike. Where once there were fumes, there is now a different kind of fuel: food.

Veg on the move

Prinzessinnengarten, in Berlin, Germany, within sight of the remaining portion of the Wall, is an urban farm with organic herbs and vegetables grown in a variety of easy-to-move containers such as boxes, rice sacks and plastic crates. Visitors can pick vegetables and learn about how to grow them. They can also just stop for a cup of coffee and admire the greenery, in what had been a place growing only graffiti. The enterprise also temporarily transforms unused urban spaces such as building sites, car parks and roofs into urban farmland.

Cichorium intybus var. *foliosum*

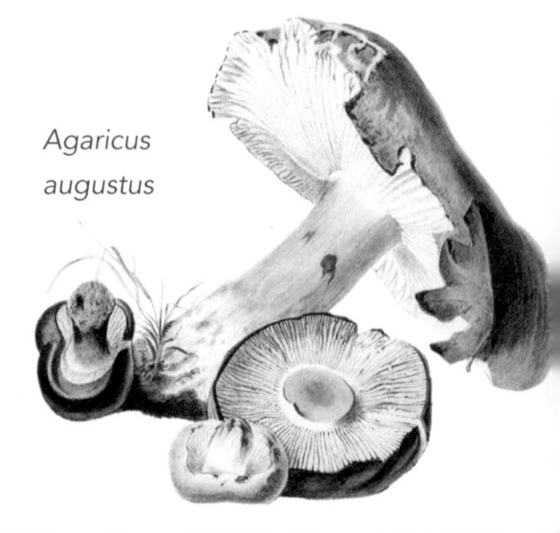

Agaricus augustus

Tunnelling for food

It's called 'vertical farming', and this particular operation occupies a Second World War air raid shelter in Clapham in south London, UK. The rumble of the underground trains can be heard in the kilometre-long shelter as employees of Zero Carbon Farms tend their crops of salad greens and herbs growing beneath the pink glow of LED lights. Crops include pea shoots, coriander, basil and rocket, but there is no soil here. Seeds are sown in carpet offcuts and then grown using hydroponic systems. The whole process, 33m (108ft) underground, is pesticide free and uses 100 per cent renewable energy. No one worries about the weather down here. Their products are delivered to stores and restaurants nearby and there are almost no food miles.

Coriandrum sativum

Other urban farms

Urban farms are everywhere, so take a look for one near you.

Beacon Food Forest, Seattle, Washington, USA: Public land turned into a 3-hectare (7-acre) edible forest garden for people to forage in.

Pasona Urban Farm, Tokyo, Japan: Company that grows vegetables and tomatoes in their office spaces, including tomatoes dangling from the ceiling.

Small spaces in São Paulo, Brazil: Agricultural technician Marcos Victorino came up with the idea of using large roof tiles, upside down, for planting vegetables and fruit on balconies, roofs and small paved areas in the city, forming a series of mini urban farms.

Urban farm in Chengdu, Sichuan, China: A 20-storey operation, built under the oversight of the Chinese Academy of Agricultural Sciences, that represents world-leading automation in farming.

Index

M

N

O

P

Q

R

S

Picture credits

B = bottom, C = centre, L = left, R = right, T = top

Cover: Rawpixel.com; Rawpixel.com/Library of Congress; Rawpixel.com/New York Public Library.

4 Rawpixel.com; 7 Rawpixel.com/Library of Congress; 8 Rawpixel.com/New York Public Library; 11 Rawpixel.com/OpenClipArt; 13 Wiki Commons/Kurt Stueber; 14 Rawpixel.com/ Wellcome Collection; 15B Rawpixel.com; 15T Rawpixel.com/Biodiversity Heritage Library; 17 Rawpixel.com/ Biodiversity Heritage Library; 19 Rawpixel.com/ Wellcome Collection; 21 Rawpixel.com; 22 Shutterstock / Dasha Si TR; 22 Shutterstock / Ermak Oksana C; 22 Shutterstock / MarBom CL; 23 Wiki Commons/ JonRichfield TR; 23 Rawpixel.com/ Biodiversity Heritage Library B; 24 Rawpixel.com; 25 Rawpixel.com; 27 Rawpixel.com/Biodiversity Heritage Library TL; 27 Rawpixel.com/New York Public Library TR; 27 Rawpixel.com/New York Public Library BL; 27 Rawpixel.com/ New York Public Library BR; 29 Cornell University; 30 Rawpixel.com/Cleveland Museum of Art; 31 Rawpixel.com B; 31 Rawpixel.com/Library of Congress T; 32 Rawpixel.com; 35 Wiki Commons/ Norton Simon Museum; 36 Rawpixel.com/Biodiversity Heritage Library; 39 Rawpixel.com/Library of Congress; 41 Rawpixel.com/ Statens Museum for Kunst; 42 Flickr / Biodiversity Heritage Library; 43 Flickr / Biodiversity Heritage Library; 44 Rawpixel.com/New York Public Library; 47 Rawpixel.com/The Rijks Museum; 48 Rawpixel.com; 51 Rawpixel.com/Princeton University Art Museum; 57 Rawpixel.com/New York Public Library; 58 Rawpixel.com/Libray of Congress; 60 Shutterstock / DragonTiger8 B; 60 Shutterstock / Evgenii Skorniakov T; 61 Shutterstock / dueanchai intaraboonsom TR; 61 Shutterstock / lady-luck BR; 61 Vecteezy.com / Matt Cole TL; 61 Vecteezy.com / Oleg Gapeenko TR; 61 Vecteezy.com / Rahim miah TL; 61 Vecteezy.com / Sonnycool TC; 64 Rawpixel.com/Library of Congress; 65 Rawpixel.com/New York Public Library; 67 Rawpixel.com; 71 Flickr / Biodiversity Heritage Library TR; 71 Rawpixel.com TL; 71 Rawpixel.com/New York Public Library BL; 72 Rawpixel.com/Biodiversity Heritage Library TC; 72 Rawpixel.com/Cleveland Museum of Art TL; 72 Rawpixel.com/Library of Congress TR; 72 Rawpixel.com/New York Public Library BR; 72 Rawpixel.com/New York Public Library CB; 75 Rawpixel.com/ Statens Museum for Kunst; 77 Rawpixel.com/ Wellcome Collection; 78 Rawpixel.com; 79 Rawpixel.com; 81 Rawpixel.com/ Cleveland Museum of Art; 82 Vecteezy.com / Oleg Gapeenko; 84 Rawpixel.com/Wellcome Collection; 85 Rawpixel.com/The British Library ; 87 Rawpixel.com/The Internet Archive B; 87 Rawpixel.com/ The MET Museum T; 88 Rawpixel.com; 90 Rawpixel.com; 92 Rawpixel.com/New York Public Library R; 92 Rawpixel.com/ The Yale University Art Gallery L; 93 Rawpixel.com L; 93 Rawpixel.com/Wellcome Collection R; 94 Rawpixel.com/ Biodiversity Heritage Library L; 94 Rawpixel.com/Library of Congress R; 95; Rawpixel.com/Biodiversity Heritage Library B; 95 Rawpixel.com/New York Public Library T; 96 Rawpixel.com/ Wellcome Collection; 97 Rawpixel.com/New York Public Library; 99 Rawpixel.com/New York Public Library; 103 Rawpixel.com/ Statens Museum for Kunst; 104 Flickr / Biodiversity Heritage Library L; 104 Rawpixel.com/Biodiversity Heritage Library R; 105 Flickr / Biodiversity Heritage Library L; 105 Wiki Commons/ Biodiversity Heritage Library R; 105 Wiki Commons/The Internet Archive C; 106 Rawpixel.com/Libray of Congress; 109 Rawpixel.com B; 109 Rawpixel.com/University of Pittsburg T; 110 Rawpixel.com T; 110 Rawpixel.com BL; 110 Rawpixel.com/Rijks Museum BR; 111 Rawpixel.com/Cleveland Museum of Art; 112 Flickr / Biodiversity Heritage Library B; 112 Rawpixel.com T; 113 Rawpixel.com T; 113 Rawpixel.com B; 115 Freepik B; 115 Vecteezy.com / Matt Cole C; 115 Vecteezy.com / Matt Cole T; 115 Vecteezy.com / Oleg Gapeenko L; 115 Vecteezy.com / Oleg Gapeenko C; 115 Vecteezy.com / Rahim miah BL; 115 Vecteezy.com / Sonnycool C; 116 Flickr / Biodiversity Heritage Library; 117 Rawpixel.com/New York Public Library; 118 Flickr / Biodiversity Heritage Library BL; 118 Flickr / Biodiversity Heritage Library T; 118 Wiki Commons/ Missouri Botanical Garden Library BR; 119 Flickr / Biodiversity Heritage Library; 120 Wiki Commons/Biodiversity Heritage Library; 124 Rawpixel.com; 126 Map by FreeVectorMaps.com; 128 Rawpixel.com/National Park Service; 129 Rawpixel.com/ The Smithsonian; 130 Rawpixel.com/New York Public Library T; 130 Rawpixel.com/New York Public Library B; 131 Rawpixel.com/ Cleveland Museum of Art; 136 Rawpixel.com/Biodiversity Heritage Library; 137 Rawpixel.com/Rijks Museum T; 137 Rawpixel.com/Rijks Museum B; 138 Rawpixel.com/The Smithsonian L; 138 Rawpixel.com/The Smithsonian R; 138 Rawpixel.com/The Smithsonian C; 139 Rawpixel.com/The Smithsonian; 141 Rawpixel.com/New York Public Library; 145 Rawpixel.com/Wellcome Collection; 147 Rawpixel.com/Biodiversity Heritage Library; 148 Map by FreeVectorMaps.com; 149 Rawpixel.com; 153 Rawpixel.com R; 153 Rawpixel.com/ Biodiversity Heritage Library L; 154 Freepik / brgfx BR; 154 Freepik / valadzionak_volha BL; 154 Shutterstock / vladmark C; 156 Rawpixel.com/Rijks Museum; 157 Rawpixel.com/ Rijks Museum; 159 Rawpixel.com/New York Public Library; 160 Rawpixel.com/Libray of Congress; 161 Rawpixel.com/New York Public Library T; 161 Rawpixel.com/New York Public Library B; 162 Rawpixel.com/The National Gallery of Art T; 162 Rawpixel.com/Wellcome Collection B; 163 Rawpixel.com T; 163 Rawpixel.com/New York Public Library B; 164 Wiki Commons/Missouri Botanical Garden Library; 166 Rawpixel.com/Biodiversity Heritage Library; 168 Rawpixel.com/ New York Public Library; 170 Wiki Commons/Biodiversity Heritage Library; 173 Rawpixel.com/ Wellcome Collection; 175 Wiki Commons/Missouri Botanical Garden Library; 176 Wiki Commons/Hortus Botanicus Leiden; 177 Rawpixel.com/New York Public Library B; 177 Rawpixel.com/ The Smithsonian T; 179 Rawpixel.com/New York Public Library; 181 Art Collection 3 / Alamy Stock Photo; 183 Wiki Commons; 185 Flickr / Biodiversity Heritage Library; 185 Rawpixel.com BL; 185 Wiki Commons/Missouri Botanical Garden Library T; 186 Flickr / Biodiversity Heritage Library; 189 Wiki Commons/ Real Jardín Botánico; 190 Rawpixel.com/Cleveland Museum of Art; 191 Rawpixel.com/Rijks Museum; 194 Rawpixel.com/Libray of Congress; 195 Rawpixel.com/New York Public Library; 196 Rawpixel.com; 198 Rawpixel.com; 199 Rawpixel.com/USDA National Agricultural Library; 200 Flickr / Biodiversity Heritage Library T; 200 Rawpixel.com/Libray of Congress B; 201 Rawpixel.com/New York Public Library; 202 Rawpixel.com; 203 Rawpixel.com/New York Public Library T; 203 Rawpixel.com/New York Public Library B; 205 Flickr / Biodiversity Heritage Library; 206 Rawpixel.com/The Smithsonian L; 206 Rawpixel.com/USDA National Agricultural Library R; 207 Rawpixel.com/New York Public Library; 208 Rawpixel.com/New York Public Library; 209 Rawpixel.com/ New York Public Library B; 209 Rawpixel.com/USDA National Agricultural Library T; 211 Rawpixel.com/New York Public Library; 212 Rawpixel.com/Biodiversity Heritage Library; 213 Rawpixel.com/New York Public Library T; 213 Rawpixel.com/USDA National Agricultural Library B; 215 Flickr / Biodiversity Heritage Library BR; 215 Rawpixel.com/Biodiversity Heritage Library BL; 217 Rawpixel.com/New York Public Library; 218 Flickr / Biodiversity Heritage Library L; 218 Rawpixel.com/Biodiversity Heritage Library R; 219 Rawpixel.com.

Every effort has been made to credit the copyright holders of the images used in this book. The publisher apologises for any unintentional omissions or errors and will insert the appropriate acknowledgement to any companies or individuals in subsequent editions of the work.

Author acknowledgements

Thanks go to Sorrel Wood and everyone at The Bright Press and Mitchell Beazley for allowing me the chance to delve into the fascinating world of plants in all its glory and for creating such a beautifully illustrated book. I am especially indebted to Izzie Hewitt and Katie Crous for their editing prowess and to the RHS for their guidance and expertise.